Estrogen Therapy. The Benefits and Risks

Frontiers of Hormone Research

Vol. 5

Series Editor
TJ. B. VAN WIMERSMA GREIDANUS, Utrecht

S. Karger · Basel · München · Paris · London · New York · Sydney

3rd International Workshop on Estrogen Therapy in Geneva, October 20–21, 1977

Estrogen Therapy

The Benefits and Risks

Sponsored by the International Health Foundation, Geneva

Volume Editors
CHRISTIAN LAURITZEN
Department of Gynaecology and Obstetrics, University of Ulm, Ulm

PIETER A. VAN KEEP
International Health Foundation, Geneva

72 figures and 45 tables, 1978

S. Karger · Basel · München · Paris · London · New York · Sydney

Frontiers of Hormone Research

Vol. 4: Melanocyte Stimulating Hormone: Control, Chemistry and Effects. Proceedings of the International Symposium on MSH, Amsterdam, July 1976. Tilders, F. J. H.; Swaab, D. F., Amsterdam, and van Wimersma Greidanus, Tj. B., Utrecht (eds.) X+244 p., 58 fig., 14 tab., 1977
ISBN 3-8055-2635-0

Cataloging in Publication
Workshop on Estrogen Therapy, 3rd, Geneva, 1977. Estrogen therapy: the benefits and risks
Volume editors: Christian Lauritzen and Pieter A. van Keep. – Basel, New York, Karger, 1978
(Frontiers of hormone research, v. 5)
Sponsored by the International Health Foundation.
1. Estrogens – adverse effects – congresses 2. Estrogens – therapeutic use – congresses 3. Menopause – congresses I. International Health Foundation II. Keep, Pieter A. van, ed. III. Lauritzen, Christian, ed. IV. Title V. Series
W1 FR946F v. 5/WP 522 W926 1977e
ISBN 3-8055-2879-5

Printed in Switzerland by Merkur AG, Langenthal
ISBN 3-8055-2879-5

Contents

Editors' Note

This is the third time that we, the editors, have had the privilege of organising an international meeting on estrogen therapy in the peri- and post-menopause. The first two meetings were concerned mainly with the beneficial effects of the therapy. On this occasion, whilst looking at the work done in the three years since our last workshop, we wanted also to consider the possible risks of the therapy, and, if possible, to weigh these against the advantageous aspects.

The meeting did not succeed in coming to a concrete conclusion in connection with this latter point for, as with most, if not all, therapies, what is right for one patient is not right for all. What the workshop did provide was a platform for the airing of views and a healthy debate on all aspects of the matter. Very often, as will be seen in the following pages, we heard 'This seems to be the case, but further studies are needed before we can be sure', but if a consensus of opinion has to be given it is probably that estrogen therapy is clearly of enormous benefit to many women, contra-indicated in some cases, and to be used with caution in others.

In publishing the proceedings of the workshop we hope to bring current opinion and up-to-date data to the attention of the prescribing physician and so to make easier his task of deciding how best to treat the individual patient.

The papers published here are those presented at the workshop, and no attempt has been made to edit them. The discussion sessions, however, had, of course, to be edited for the sake of clarity and brevity, but all the speakers have had a chance to read a draft of the manuscript and to ensure that the comments printed here are as they wish them to appear.

We are extremely grateful to all present at the workshop, especially to those who read papers, for sharing with us their latest findings and opinions, and our special thanks are due to Professor Nordin who so ably chaired the general discussion session, and upon whom fell the unenviable task of trying to balance the benefits of the therapy against the risks.

Finally, we acknowledge with grateful thanks, the enthusiastic help of Pamela Freebody – now for the third time – with the editing task.

Christian Lauritzen
Pieter A. van Keep

List of Participants

President

Dr. H. KOPERA, Department of Experimental and Clinical Pharmacology, University of Graz, Universitätsplatz 4, *A-8010 Graz,* Austria

Chairmen

Prof. C. LAURITZEN, Department of Gynaecology and Obstetrics, Universität Ulm, Frauenklinik, Prittwitzstrasse 43, *D-7900 Ulm/Donau,* FRG

Dr. P. A. VAN KEEP, International Health Foundation, Case postale 761, *CH-1211 Geneva 1,* Switzerland

Participants

Prof. E. E. BAULIEU, Département de Chimie Biologique, Faculté de Médecine de Bicêtre, Université de Paris Sud, 78 avenue du Général Leclerc, *F-94 Bicêtre,* France

Dr. P. C. BRAND, International Health Foundation, Case postale 761, *CH-1211 Geneva 1,* Switzerland

Prof. A. C. COMNINOS, Department of Obstetrics and Gynaecology, University of Athens, 20 Kanari Street, *Athens 138,* Greece

Dr. C. VAN DER DOES, Department of Obstetrics and Gynaecology, University Hospital Leiden, Rijnsburgerweg 10, *Leiden,* The Netherlands

Prof. A. A. HASPELS, Department of Gynaecology and Obstetrics, University Hospital Utrecht, Catharijnesingel 101, *Utrecht,* The Netherlands

Prof. P. J. KELLER, Leiter der Abteilung für Endokrinologie, Universität Zürich, *CH-8091 Zürich,* Switzerland

Prof. A. KLOPPER, Department of Obstetrics and Gynaecology, University of Aberdeen Medical School, Foresterhill, *Aberdeen AB9 2ZD,* United Kingdom

Prof. H. M. Lemon, Section of Oncology, Department of Medicine, University of Nebraska Medical Center, 42nd Street and Dewey Avenue, *Omaha, NB 68105,* USA

Dr. T. M. Mack, Director, Cancer Surveillance Program, Department of Pathology, University of Southern California School of Medicine, 2025 Zonal Avenue, *Los Angeles, CA 90033,* USA

Dr. D. McKay Hart, Division of Obstetrics and Gynaecology, Stobhill General Hospital, *Glasgow G21 3UW,* United Kingdom

Prof. E. Myhre, Institutt for Patologi, Rikshospitalet, *Oslo 1,* Norway

Prof. M. Neves-e-Castro, Medizinische Klinik C, Universität Düsseldorf, *Düsseldorf,* FRG

Prof. B. E. C. Nordin, MRC Mineral Metabolism Unit, The General Infirmary, Great George Street, *Leeds LS1 3EX,* United Kingdom

Prof. A. E. Rakoff, Department of Obstetrics and Gynecology, Jefferson Medical College of Thomas Jefferson University, *Philadelphia, PA 19107,* USA

Prof. L. Rauramo, Department of Obstetrics and Gynaecology, Turku University Central Hospital, Hämeenkatu 4, *20520 Turku 52,* Finland

Dr. T. E. Shellenberger, USA Department of Health, Education and Welfare, Division of Molecular Biology, National Center for Toxicological Research, *Jefferson, AR 72079,* USA

Dr. J. W. W. Studd, Department of Obstetrics and Gynaecology, King's College Hospital, Denmark Hill, *London SE5 9RS,* United Kingdom

Dr. J. H. H. Thijssen, Department of Endocrinology, University Hospital Utrecht, Catharijnesingel 101, *Utrecht,* The Netherlands

Dr. W. H. Utian, Department of Reproductive Biology, Case Western Reserve University School of Medicine, *Cleveland, OH 44106,* USA

Prof. K. D. Voigt, 11. Medizinische Klinik, Abt. für Klinische Chemie, Universität Hamburg, Universitäts-Krankenhaus Eppendorf, Martinistrasse 52, *2 Hamburg 20,* FRG

Prof. G. P. Vooijs, Instituut voor Pathologische Anatomie, Catholic University of Nijmegen, Geert Grooteplein Zuid 24, *Nijmegen,* The Netherlands

Prof. F. de Waard, Institute of Social Medicine, University of Utrecht, Bijlhouwerstraat 6, *Utrecht,* The Netherlands

Dr. Tj. B. van Wimersma Greidanus, Rudolf Magnus Institute for Pharmacology, University of Utrecht, Vondellaan 6, *Utrecht,* The Netherlands

Welcoming Address

It is a great privilege for me to welcome, on behalf of the International Health Foundation, so many distinguished scientists to the third of our workshops on menopause research. In the previous two what might be called the basic physiological, endocrinological, and therapeutic aspects of 'ageing and estrogens' and 'estrogens in the post-menopause' were discussed. This one will be devoted to what is of paramount and immediate importance to the patient in this age group and to her therapist, namely, the pros and cons of estrogen therapy.

A critical review of the prevailing opinions and, perhaps, the formulation of recommendations concerning this subject are urgently needed. I am certain that the proceedings resulting from this meeting will contain not only most valuable reports on research activities, but also well-considered authoritative opinions, and many useful suggestions of direct practical importance regarding the therapeutic use of estrogens.

One may wonder why, when the physiological event of the menopause and all its accompanying problems must have existed since the beginning of mankind, so much scientific and public interest should only now be devoted to the period around the menopause and the time thereafter. It seems that mere existence is not the thing that matters. The importance of facts changes with time, and this holds true for the climacteric. What then has brought about these claimed changes? To my mind there are, among others, three influences of major importance.

Firstly, there is prolongation of life expectancy. The fact that for women in industrialized countries this is now some 70 or 75 years means that most can now expect to spend one third of their lives in the post-menopause. Consequently, there is now much more time than before in which to notice, experience, and be bothered by, climacteric and post-climacteric changes. The second influencing factor is the emancipation of the woman. Our mothers and grandmothers usually spent their lives for and with the family. Only in very rare instances were women actively engaged in the working process of professional life. The woman's place was in her home, and it was her intention that this should be so. Her main responsibility was the household, and the care of her husband and children and of the aged of the family. In such surroundings the experience of – let us say – hot flushes and other unpleasant climacteric complaints was of relative unimportance, and so was their prevention. This situation has drastically changed. Nowadays, when a woman is incommoded by a burst of sweating or a red face, by an unexplainable attack of sadness, and so forth, while in the office or in a place where others watch her and expect optimal efficiency from her, such symptoms become very real problems. In our time, therefore, women have become increasingly concerned about their well-being, at any rate in the period around the time of the menopause. Thirdly, modern medicine has much more than mere moral support to offer to sufferers from ailments connected with progressing ovarian insufficiency. Intensive research has resulted in the development of a number of therapeutic possibilities, above all, that of estrogen replacement.

Optimal therapy always depends on factors such as sufficient knowledge of the cause of a troublesome condition, of the actions of the therapeutic agent, and of the benefit/risk balance. It is my sincere hope that the next two days will contribute substantially to the enlargement of this knowledge, to the benefit of all women and their families.

Dr. Hans Kopera
Secretary of the Board of Trustees
of the International Health Foundation

Front. Hormone Res., vol. 5, pp. 1–25 (Karger, Basel 1978)

Proven Beneficial Effects of Estrogen Substitution in the Post-Menopause – a Review

C. LAURITZEN and P. A. VAN KEEP

Department of Gynaecology and Obstetrics, University of Ulm, Ulm
and International Health Foundation, Geneva

In every kind of treatment the expected beneficial effects of the therapy have to be weighed critically against possible risks and disadvantages. To make this feasible, drug actions and body reactions must be exactly defined and assessed. Modern therapeutic studies have proved that alleged drug actions may be attributed, in addition to real pharmacodynamic effects, to placebo effects and to the influence of suggestions made by the prescribing physician to the patient. The physician himself is also subject to autosuggestion, resulting from his wishes, his theoretical views, and even his prejudices. It is therefore essential for the scientific evaluation of the therapeutic value of drugs, including hormones, to take into account only those actions which have been proven beyond any doubt to be caused by the medicament and which can be expected, with a reasonable degree of certainty, to take place in the patient. These actions should then be recorded without putting suggestive questions to the patient.

Regrettably there is so far no exact quantitative method to measure climacteric complaints. Certainly the Blatt and the Kupperman menopausal indeces contain too many unspecific symptoms, which are all summed up together. The best method still seems to us to be the simple counting of hot flushes per time unit. The effectiveness of the therapy should preferably have been proved by means of a double-blind, cross-over evaluation against a placebo. The sum of these proven beneficial effects, minus the sum of possible disadvantageous effects, would ideally constitute the therapeutic value of a drug. This value would be the greater the more the advantages predominate and the more accurately the positive and negative effects of medication are known beforehand.

It should not be forgotten, however, that therapeutic trials are by no means exact physical experiments, but that they are approximations only. The experiences and the convictions of the clinician and his judgement of the success rate of estrogen treatment in his patients, who are well-known to him, should also be taken into consideration. Lastly, the opinion of the patient herself as to whether the treatment has brought real relief to her should not be underrated or disregarded. In our opinion the evaluation by the patient definitely has a place in the judgement of a drug. What finally matters to the woman is not the objective therapeutic action, but the subjectively experienced improvement of her climacteric complaints.

The purpose of this paper is to analyse which of the climacteric symptoms, as complained of by the patients and as described by different authors, are firmly attributable to the decrease or cessation of ovarian hormone production and which are not, and to review to what extent estrogen therapy has been proved to prevent or cure the changes and symptoms occurring around the time of the climacteric, or caused by it. The possible disadvantages of estrogen therapy are not discussed; this will be done in other contributions to this volume.

Typical Climacteric Complaints

There are no *specific* climacteric complaints. All symptoms which occur in the climacteric female can also occur in other, non-climacteric, age groups and in connection with other disturbances or diseases. They can, therefore, be caused by factors other than estrogen deficiency. *Typical* climacteric complaints, however, are those which begin when estrogen production declines or stops, and which occur either spontaneously in the perimenopausal years or enforced after bilateral oophorectomy. These typical climacteric complaints are most likely caused directly or indirectly by estrogen withdrawal and they can be treated very successfully with estrogen substitution. Treatment with a placebo would be much less effective [WIED, 1953; GREENBLATT *et al.*, 1950; LAURITZEN, 1973; UTIAN, 1975, 1977; CAMPBELL, 1976; CAMPBELL and WHITEHEAD, 1977; COOPE, 1976]. Typical climacteric complaints in this sense are only the vasomotor complaints, namely hot flushes and profuse sweating, often combined with dizziness and a tingling sensation in the upper extremities [NOVAK, 1954; LAURITZEN, 1973; UTIAN, 1975]. However, some ill-defined deterioration of the psychic state, of drive and motivation, seems also typical for estrogen deficiency. Clear placebo effects are found for the dry vagina, coital dissatisfaction, pollakisuria and headaches [CAMPBELL, 1976]. Palpitations, sometimes occurring at the same time as hot flushes, are apparently not directly caused by estrogen deficiency [UTIAN, 1973], but are a symptom resulting from anxiety or from a temporary secondary disturbance of the autonomous nervous system on a constitutional basis. Insomnia, which is a common complaint, may exist partly because of the occurrence of hot flushes and profuse sweating at night, but may also be caused by feelings of depression and by tensions resulting from the disturbance of the patient's relations with her surroundings. Indeed in most cases insomnia improves very rapidly after the flushes have disappeared. Depression and other psychic symptoms disappear only after the fourth or fifth treatment cycle [MOORE *et al.,* 1975]. Nervous and mental complaints such as depressions, feelings of decreased vitality and drive, of hopelessness, of diminished mental capacity and tolerance, of restlessness, nervousness and irritability often predominate in the climacteric picture. Fatigue is a rather common complaint too. The loss of psychic vitality may be caused directly by estrogen withdrawal, since there is a specific effect of estrogen treatment on the feeling of well-being [UTIAN, 1975]. Most psychic symptoms, however, are probably caused by anxiety resulting from the debilitating complaints, by occult depression, and also by the changing personal and social situation with which the patient is confronted. Autonomic nervous symptoms, neuroses or endogenous depressions, which until then may have been compensated, may now also become manifest as a result of the current crisis in the patient's life. The symptomatology may begin in the premenopause. The maximum of complaints usually occurs 2–3 years after the menopause, and then slowly decreases [NEUGARTEN and KRAINES, 1966; JASZMANN, 1973, 1976; MCKINLAY and JEFFERIES, 1974; DEWHURST, 1976].

Late Consequences of Estrogen Deprivation

One direct, though late, consequence of estrogen loss is atrophy of the target organs, namely the vulva, the vagina, the urethra and the bladder, in particular its trigone. Simple vulval atrophy, sometimes accompanied by pruritus or burning, is certainly a sign of lack of estrogen [HAUSER *et al.,* 1961; LAURITZEN, 1972; BEARD, 1976]. Lichen sclerosus et atrophicus, however, is a localized disease which can also occur at other ages without atrophy and in other locations [LABHART, 1955; COCKERELL and KNOX, 1969; DAPUNT *et al.*, 1971]. In aged women, however, the additional local vulval atrophy is an important constituent of the disease [LABHART, 1955]. Atrophic vaginitis with kraurosis vaginae, and sometimes discharge of blood, and dyspareunia or vaginal dryness, is clearly caused by estrogen deficiency [CAMPBELL, 1976]. This is also true of the simple urethral caruncle (urethral ectropium), which often occurs in this age group and which is sometimes misdiagnosed as a polyp. Atrophic cystitis can be caused by estrogen deficiency and may lead to urge incontinence. Other factors can of course be of additional causal significance. Infection and inflammation of the bladder may be superimposed on atrophic involution.

Descensus uteri et vaginae and urinary stress incontinence are conditioned by many factors, such as constitution, number and mode of deliveries, as well as strenuous physical activity. Estrogen deprivation, however, is obviously an additional important cause, because a large percentage of cases of descensus, prolapse and of urinary incontinence manifest themselves at the period of the climacteric. Slackening of the pelvic floor may be caused by atrophy and weakening of the muscular, ligamental and connective tissue support of the genital organs as a result of estrogen deficiency [LABHART, 1955; KROKEFORS, 1960; LAURITZEN and MUELLER, 1977]. Atrophy of the breast is also a frequent occurrence in post-menopausal women [LIEPELT, 1952].

Osteoporosis

This disease is five times more frequent among old women than among men. It may be said that amongst old people suffering from osteoporosis, 85% are women and 15% are men. About 25% of women over 60 years have spinal compression fractures. 75% of all fractures in women over 45 years and 80% of hip fractures occur in osteopenic bones. Within 3 months of hip fracture, the mortality is 16% [ALFRAM, 1964; KNOWELDEN *et al.*, 1964; HEANEY, 1974; HEANEY and RECKER, 1975]. It is not possible to predict which female patients will develop osteoporosis in the postmenopause, and it is not known why this disease occurs in one individual and not in another. It may be related to the constitutional bone mass, to the intake and resorption of calcium and vitamin D, to lack of exercise or to other factors. However, in the climacteric woman estrogen deprivation is certainly an important factor in the development of osteoporosis [NORDIN *et al.*, 1975]. After castration, the first loss of bone from the cortex occurs within a few years [AITKEN, 1976a, b]. Numerous biochemical investigations have shown that estrogens probably inhibit the activity of the osteoclasts and may reduce a negative calcium balance by inhibiting the loss of calcium via urine and faeces [GALLAGHER and NORDIN, 1973, 1975; DEQUEKER *et al.*, 1975]. It has been demonstrated that in early castrated female patients there is a greater loss of bone and more fractures of the radius and the neck of the femur in old age than in a normal population [JOHANSSON *et al.*, 1975].

Lipid Metabolism, Atherosclerosis, Myocardial Infarction

Following natural menopause or after castration there is an increase in total lipids and cholesterol, especially in the low density lipoprotein fraction. Phospholipids and triglycerides show no significant increase [BOYD, 1973; PUNNONEN and RAURAMO, 1976]. An elevated cholesterol level is known to be an important risk factor for the development of atherosclerosis and the occurrence of myocardial infarction; this has been proven by the Framingham Study and by various authors [FURMAN, 1969]. The incidence of coronary heart disease increases after the menopause [KANNEL *et al.*, 1976]. In women who lose their ovaries at an early age it has been shown that there is a significant increase of angina pectoris and myocardial infarction in the post-menopause [OLIVER and BOYD, 1959; LONDON *et al.*, 1961; MANCHESTER *et al.*, 1971; OLIVER, 1974; JOHANSSON *et al.*, 1975]. Most retrospective studies by clinicians and pathologists show this increase and suggest a role of ovarian estrogen in the incidence of atherosclerosis, coronary sclerosis and myocardial infarction [WUEST *et al.*, 1953; RIVIN and DIMITROFF, 1954; ASK-UPMARK, 1962; HIGANO *et al.*, 1963; SZNAJDERMAN and OLIVER, 1963; BERKSON *et al.*, 1964; PARRISH *et al.*, 1967; review: BEARD, 1976]. However, there are also studies which could not show such a difference in the incidence of coronary diseases [NOVAK and WILLIAMS, 1960; RITTERBAND *et al.*, 1963]. It must, however, be pointed out that in the study of RITTERBAND *et al.* [1963] the frequency of myocardial infarction was also rather high in the comparison group. Women with ischaemic heart disease have been found more often than controls to have had an early menopause [SPITZER *et al.*, 1957; BENGTSSON, 1973; KANNEL *et al.*, 1976]. BLANC [1977] found no correlation between the age at menopause and the age at which myocardial infarction occurs, but no firm conclusion can be drawn from this study of 94 patients with a natural menopause and 10 castrated patients with a mean age of 47.2 years.

Thus much of circumstantial evidence points to a protective effect of endogenous estrogens against atherosclerosis, especially coronary sclerosis and myocardial infarction. However, further, more clear and convincing data have to be produced before the relative significance of the factor estrogen can be quantitatively determined.

Hypertension

Systolic and diastolic blood pressure increases temporarily when hot flushes occur. Most epidemiologic studies show that in post-menopausal women the mean systolic and diastolic blood pressure increases more often and higher than it does in men of the same age group [BUERGER, 1958; PICKERING, 1955]. Two thirds of all patients with essential hypertension are women between 45 and 70 years of age, and the disease usually begins at the time of the climacteric [BENEDICT, 1956; PRILL and LAURITZEN, 1970]. Within the group of hypertensive patients, women have higher blood pressure levels than men in the large majority of cases. About 15 % of all women in the age group 45–55 have a systolic blood pressure of more than 160 mm Hg. Women with severe climacteric complaints significantly more often have hypertension than do women without climacteric complaints [PRILL, 1966]. It has been hypothesized that the loss of estrogens might be partly responsible for the mean statistical rise in blood pressure in post-menopausal women. There seems, in fact, to be a protective effect of parenteral estradiol valerate on reactive blood pressure elevation caused by stress [VON EIFF, 1975]. However, any increase in blood pressure is a multifactorial phenomenon, and the role of the one factor – estrogen deficiency – in this connection is not very well known and has to be made clear by further investigations.

Weight Increase

Obesity after castration is well known from animal experiments. There is a weight increase of more than 2.5 kg in about half of all women 3–5 years after the menopause (American life insurance statistics, 1970). This increase may be brought about by many other causes such as metabolic-endocrine disturbances, hypothyroidism, or by hyperphagia, due to oral compensation of psychic frustration at the time of the climacteric estrogen loss. Metabolic balance studies with modern methods have not yet been published for post-menopausal patients and there is an urgent need for prospective investigations in this field. In conclusion, the relationship of post-menopausal estrogen deprivation to body weight in the human female is at present not clear.

Skin

A few years after the menopause, and only some months after castration, skin atrophy occurs. This is experienced by the patient as thinning, increased dryness, loss of elasticity and wrinkling of the skin. These changes primarily result from a reduced mitotic activity in the stratum germinativum of the epidermis [Rauramo and Punnonen, 1973; Shahrad and Marks, 1976]. The elastic and the subcutaneous connective tissue also undergo degenerative changes. Loss of estrogens leads to a decrease in the retention of sodium and water in the extracellular space and to a reduced production of hyaluronic acid, which also causes a loss of water binding. There is a diminution of blood perfusion and the skin temperature decreases. Hair growth also changes. Pubic hair and sometimes scalp hair becomes thinner. Some post-menopausal patients (about 30%) develop hirsutism [Berblinger, 1925; Lauritzen and Mueller, 1977]. This is partly due to the disappearance of the natural, antiandrogenic action of estrogens and to the continuing production of androgens by the adrenals and sometimes by the ovaries. The voice may also show a reduction in the upper register and a loss of timbre. Dryness of nose, ears and eye membranes may also occur [Prijot *et al.*, 1972].

Breasts

In about 60% of all women in the post-menopause a decrease occurs in the volume and firmness of the breast. The glandular tissue is substituted by fat and connective tissue [MacLaren, 1941; Liepelt, 1952]. The decreased firmness of the breasts may be connected with estrogen-progestagen deficiency. Age-related tissue changes also play a role.

Rheumatoid Arthritis and Osteoarthritis

A significant increase in periarticular complaints has been noted in post-menopausal women [Greenblatt, 1963; Dequeker *et al.*, 1977] and a clear correlation of such 'rheumatoid symptoms' with the severity of climacteric symptoms has been found [Prill and Lauritzen, 1970]. Thus, estrogen withdrawal may play a part in the causation of these complaints. Whether this is due to a decrease of blood perfusion or by an alteration in tissue metabolism is not clear at present.

Proven Beneficial Effects of Estrogen Substitution

Few therapeutic regimens have been so thoroughly investigated as has estrogen therapy. The beneficial effects of the different estrogens are certainly better known and understood than those of many other drugs.

Vasomotor Symptoms

Hot flushes and profuse sweating are quickly and significantly reduced by estrogen treatment [GREENBLATT *et al.*, 1950; LAURITZEN, 1973; UTIAN, 1975; COOPE, 1976]. The hot flush is apparently the most typical and most sensitive clinical indicator of the effectiveness of estrogens. In some patients with hot flushes a placebo may produce some improvement for a short time. It is, however, always necessary to give estrogen to abolish the flushes completely [CAMPBELL and WHITEHEAD, 1977]. After a change from estrogen to placebo in cross-over designed studies the hot flushes worsen considerably. After changing from placebo to estrogens the condition is significantly improved [COOPE, 1976]. Estrogen medication also abolishes dizziness and tingling sensations and the concomitant rise of blood pressure which occurs with hot flushes [LAURITZEN and VELIBESE, 1961]. The sweating which occurs in connection with hot flushes is better controlled by estrogens than by a placebo. After crossing over to placebo therapy from estrogen treatment there is a worsening of this symptom [LAURITZEN, 1973].

Nervous and Psychic Complaints

The mental tonic effect of estrogens is proven [GREENBLATT, 1965; BOLDING and WILLCUTT, 1969; UTIAN, 1972; HUNTER *et al.*, 1973]. In double-blind studies depressive moods or melancholia show a slight improvement on placebo therapy. This improvement is much more pronounced, however, on estrogen treatment. After changing from estrogen to placebo there is a deterioration in the melancholia. On the other hand, after changing from placebo to estrogen there is a significant improvement of the depressive mood state [COOPE, 1976; CAMPBELL and WHITEHEAD, 1977]. Memory, good spirits and optimism are also improved. Emotional functions seem to be more favourably influenced than cognitive variables [RAURAMO *et al.*, 1975]. Anxiety and worry about oneself often disappear [CAMPBELL and WHITEHEAD, 1977]. KANTOR *et al.* [1968] and MICHAEL *et al.* [1970] in double-blind studies on geriatric patients showed that estrogen therapy over more than 36 months improved learning ability and concentration. During placebo treatment of the same duration these parameters worsened. Estrogens also improve the readiness for social contact and psychic alertness in aged women [CALDWELL, 1952; DUEKER,

1957; EVANS and MARMORSTON, 1963; CALDWELL and WATSON, 1954; KOPERA, 1973]. FEDOR-FREYBERGH [1977] demonstrated in a double-blind trial that estrogen medication exerts a positive influence on many psychological functions in the climacteric. Anxiety, fatigue, restlessness, irritability, depression, disturbances in concentration, loss of memory, and sleeplessness (apparently caused by the primary climacteric symptoms hot flushes and sweating) are reduced. Improvement of memory was also found by CAMPBELL [1976]. Estriol increases attention and alertness [VANHULLE and DEMOL, 1976]. Headaches are not a climacteric symptom; they are seldom improved and can, in some cases, even be worsened by high doses of estrogen. Insomnia and arthralgia respond better to estrogen than to placebo therapy [COOPE, 1976]. Insomnia and anxiety react more favourably to tranquillizers than to estrogens; they react, however, best to a combination of both [SHEFFERY *et al.,* 1969]. The psychotrophic effect of estrogens can also be objectivated by biochemical methods such as monoamine oxidase activity [KLAIBER *et al.,* 1976] and free plasma tryptophan [AYLWARD, 1975].

Atrophy of Target Organs

Vulval atrophy and vulvitis based on atrophy are consistently and safely improved by estrogen, administered either locally or parenterally [LAURITZEN, 1970; RAURAMO, 1976]. The same holds for atrophic colpitis and kraurosis vaginae [BORGLIN, 1959; GITSCH and GOLOB, 1962; GAD *et al.,* 1971; KORTE, 1973; UTIAN, 1975]. Vaginal dryness may also be improved by a placebo, but to a lesser degree [CAMPBELL, 1976]. Lichen sclerosus et atrophicus in the climacteric, especially when based on vulval atrophy, is favourably influenced by the local application of estrogens and corticosteroids or by the oral or parenteral administration of estrogens and even androgens over a longer period of time [DAPUNT *et al.,* 1971]. A consistent and favourable effect is also obtained by local and parenteral estrogen administration in atrophic conditions, such as vaginitis and kraurosis vulvae et vaginae, when painful intercourse (dyspareunia) is the problem [JOSWIG-PRIEWE *et al.,* 1973; LAURITZEN and MUELLER, 1977].

Urethral caruncle (ectropium urethrae) reacts very readily to estrogens, disappearing completely after a few weeks of estrogen treatment [HOFFMANN, 1950a, b; LAURITZEN and MUELLER, 1977]. Mild and medium degrees of utero-vaginal descensus have occasionally been reported to improve with estrogen medication [ANDERSON, 1958]. Atrophic cystitis and urethritis, with corresponding symptoms, can certainly be ameliorated by estrogen medication as shown by urinary cytology and disappearance of the complaints [HOFFMANN, 1950a, b; LAURITZEN, 1968; JÖNSSON, 1973; SMITH, 1977]. There are also clinical reports of an estrogen-induced improvement of radiogenic colpitis [PITKIN and BRADBURY, 1965] and of urge incontinence [RAURAMO, 1976]. Estriol, ethinyl estradiol, cyclopentylenolether and estradiol valerate have been reported to improve first and even second degree stress incontinence [LAMBILLON, 1971; WALLNER and SOOST, 1971; BOL, 1960; KRONIG, 1959; SLUNSKY, 1963; CAINE and RAZ, 1973; DAXELMUELLER *et al.,* 1974]. None of these studies, however, has been double-blind. Some investigations have demonstrated an increase in maximal urethral pressure during estrogen treatment. Yet, in many cases where the urethral pressure did not change, a clinical improvement was seen. Only a few cases showed no change at all [SMITH, 1976, 1977; HARRISON, 1976]. One recent double-blind study with a placebo has shown estrogens to improve urinary stress incontinence [WALTER *et al.,* 1976]. Thus in our opinion there is at least some evidence that estrogens may improve stress incontinence. Androgens and progestagens (nortestosterone derivates) are apparently also effective [FASTNER, 1975].

Disturbances of Libido and Orgasm

Libido decreases in a considerable percentage of cases after the menopause [KINSEY *et al.,* 1953; HÄLLSTRÖM, 1977; PRILL and LAURITZEN, 1970; PFEIFFER and DAVIS, 1972; VAN KEEP and GREGORY, 1977]. Estrogens as such have no influence on libido and orgasm. There is a behavioural fixation of sexual attitudes in the mature woman. However, in some climacteric women libido is decreased because of nervousness, irritability and depressive mood, severe hot flushes and a feeling of ill health. In such cases libido and sexual activity as well as satisfaction, experience of pleasure, sexual phantasies, orgasmic capacity and 'sexual relevance' can be improved by estrogen treatment [FEDOR-FREYBERGH, 1977]. When vulval or vaginal atrophy and inflammation lead to dyspareunia, the treating of these symptoms with estrogens may also restore normal sexual feelings and activity by normalizing the structure and function of the organs [MASTERS and JOHNSON, 1966]. Estrogens certainly improve vaginal lubrication [MONEY and DALÉRY, 1977]. There is a true stimulating effect of testosterone on libido and sexual drive [GREENBLATT, 1943, 1963, 1965; KUPPERMAN *et al.,* 1953; MONEY, 1961; BIEREN, 1963; KUPPERMAN, 1967; JERN, 1973; STUDD *et al.,* 1977] and this may be an indication for the addition of androgen when treating a decrease of libido and anorgasmy.

Metabolic Consequences

Osteoporosis

Numerous investigations have shown that estrogens inhibit the activity of osteoclasts, decrease calcium levels in plasma, and normalize a negative calcium balance by reducing calcium loss through urine and faeces [e. g., RIGGS *et al.,* 1972; AITKEN *et al.,* 1973a, b, 1974]. In retrospective clinical studies it has been demonstrated that the prophylactic administration of estrogens retards the loss of bone density [DAVIS *et al.,* 1970; DEQUEKER *et al.,* 1975; AITKEN, 1976a, b; NORDIN, 1976] and prevents a loss of height caused by osteoporotic kyphosis and vertebral crush fractures [WALLACH and HENNEMANN, 1959; HERRNBERG, 1960; NORDIN, 1976]. The number of forearm and probably hip fractures can also be reduced by long-term estrogen intake [GORDAN, 1977]. There are now prospective studies from MEEMA *et al.* [1975], DEQUEKER and FERIN [1977], FURUHJELM [1976], LINDSAY [1976], NORDIN [1976] and AITKEN [1976a, b], demonstrating that in non-substituted castrated women the bone loss is significantly greater (as measured by the change in cortical thickness) than in patients treated with ethinyl estradiol or mestranol in a daily dose of 20–25 μg. At an FDA meeting in February 1977, held to review the indications for estrogen use, the indication 'prophylaxis against osteoporosis' was approved because of the convincing evidence presented.

Osteoporosis is certainly a disease of medico-social importance. However, although fractures are relatively common in older women, post-menopausal women are not certain to get them if they do not take estrogens. When talking of the prophylactic long-term administration of estrogen, one is probably talking of a benefit to a rather small number of individuals, possibly only 10–15 %. It would, therefore, be highly desirable to develop methods to determine which patients are at risk for osteoporosis and bone fractures. Nevertheless, the prophylactic and therapeutic effects of estrogens in osteoporosis are certainly a major benefit of estrogen substitution.

Rheumatoid Arthritis and Osteoarthritis

Estrogen therapy alleviates the periarticular complaints in post-menopausal patients in a high percentage of cases [PRILL and LAURITZEN, 1970]. Double-blind studies have not been done. The mechanism of the action of estrogens in this syndrome is not clear.

Skin

The atrophic changes which occur in the skin following oophorectomy can be prevented and even reversed by the administration of estrogen. Estriol succinate and estradiol valerate have been used in the usual substitution doses [RAURAMO and PUNNONEN, 1973; PUNNONEN and RAURAMO, 1976]. The anti-atrophic action of estrogens is demonstrated by epidermal thickness measurements and by thymidine incorporation rate determinations. These findings have been confirmed to some extent by the radiological measurement of dermal thickness and by measuring the incorporation of thymidine and proline into human skin explants in the presence of estrone [SHAHRAD and MARKS, 1976; MARKS and SHAHRAD, 1977]. From the clinical point of view, the effect of estrogens on dermal thickness, skin appearance, wrinkles, elasticity and blood perfusion has been studied by AERTGEERTS [1972]. This author has found an improvement of these parameters after 6 and 12 months' estriol succinate treatment (4 mg) in a high percentage of patients. The improvement of skin quality by the application of estrogen-containing ointment had already been shown [GOLDZIEHER, 1946; CHIEFI, 1950; POCKRANDT, 1954; BROWN, 1966].

Breasts

The increase in breast volume after administration of estrogens and of an estrogen-progestagen combination is well-known from experience with hormonal oral contraceptives [PINCUS, 1965]. Presumably the volume and the glandular and connective tissue of the breast can be improved in ageing women by estrogen-progestagen medication. The simultaneous administration of progestagens with the estrogen will prevent possible risks and in some cases even improve mastopathia fibrosa cystica [JERN, 1973; Royal College of General Practitioners, 1974]. However, methods of a more exact assessment of estrogen-progestagen effects on the breasts are needed.

Questionable Benefits of Estrogen Substitution

Lipids

Administration of various estrogens, such as ethinyl estradiol, mestranol, conjugated estrogens, micronized estradiol, estradiol valerate and estrone sulphate nearly always produces a decrease of total lipids and of total cholesterol, especially of β-lipoprotein cholesterin, while increasing the phospholipids [BOYD, 1973]. Ethinyl estradiol, mestranol and conjugated estrogens (in doses of more than 2.5 mg) usually elevate triglyceride levels [FURMAN, 1969; LEBECH and BORGGAARD, 1973; BOYD, 1973]. Estradiol valerate, micronized estradiol and estriol, in the usual therapeutic dosages, however, do not increase the triglycerides. Estriol and estradiol may sometimes even decrease triglyceride levels if they are elevated [PROOST *et al.*, 1969; LARSSON-COHN, 1976; PUNNONEN and RAURAMO, 1976]. A high cholesterol level is a risk factor for atherosclerosis. Lowering the cholesterol is, therefore, thought to be advantageous in connection with the prevention of atherosclerosis, but this has not been proven beyond doubt. Since atherosclerosis and myocardial infarction are multicausal diseases, it is difficult to assess the significance of one isolated causal factor.

In retrospective studies the expected frequency of coronary thrombosis has been seen to be significantly lower in women on long-term estrogen treatment than in non-treated populations of similar age and residence [POTOCKI, 1971; BURCH *et al.*, 1974]. In the epidemiological study of ROSENBERG *et al.* [1976] the relative risk of myocardial infarction in long-term estrogen-treated patients was slightly lower (relative risk 0.97) than in the comparison group. These investigations do not contradict a possible prophylactic effect of estrogens. They are, however, by no means to be considered as positive proof. PFEFFER and VAN DEN NORT [1976] found that post-menopausal women treated with estrogens were not protected from cerebrovascular disease. So far, a prophylactic effect of estrogens on the development of atherosclerosis cannot be listed under proven beneficial effects.

Blood Pressure

Estrogen treatment abolishes the short temporary increase in blood pressure which accompanies hot flushes [LAURITZEN and VELIBESE, 1961]. Estradiol valerate intramuscularly exerts a protective effect on blood pressure, especially when the hypertension is caused by mental stress [VON EIFF *et al.*, 1975].

In early castrated patients there is a significant increase in cardiovascular morbidity and mortality 30–65 years after the operation [JOHANSSON *et al.*, 1975]. According to DAVIS *et al.* [1961] prophylactic estrogen medication in the post-menopause after castration reduces the incidence of hypertension. This only becomes apparent after an interval of more than 12 years. The statistical data of this investigation are, however, too small to be convincing.

Administration of 2–3 times the substitution doses (4 mg estradiol valerate or 5 mg conjugated estrogens) to 50 castrated women over a period of 1 year influenced the diastolic blood pressure in only 2 patients [UTIAN, 1976, 1977]. A few cases have been described where ethinyl estradiol, mestranol and also conjugated estrogens may have increased the blood pressure, apparently via their hormonal action on the angiotensin-renin-aldosterone system [e. g. CRANE and HARRIS, 1969; SPELLACY and BIRK, 1972; UTIAN, 1975; COOPE, 1976]. In any case, estriol does not increase the blood pressure [RAURAMO, 1975; STROHFELD, 1977].

There is a lack of good evidence that the long-term administration of estrogens has a favourable influence on blood pressure. In cases where there is no hereditary predisposition to hypertension there is apparently also no negative influence of estrogens on blood pressure. Prospective studies are needed on this.

Weight Increase

Most papers dealing with this subject have demonstrated that no significant weight increase occurs following therapy with conjugated estrogens or estradiol valerate [MCEVEN, 1965; UTIAN, 1972; LACHNIT, 1970; NOTELOVITZ, 1975; STROHFELD, 1977; UTIAN, 1977], neither has any weight increase been seen during estriol medication [PROOST, 1969; CATTOOR *et al.*, 1970]. We [LAURITZEN, 1973] have shown that in two randomized groups of post-menopausal patients, one with and one without estrogen therapy, there was an initial slight mean weight increase in the group which received conjugated estrogens, presumably because of water retention. However, after some months the body weight tended to return to the level before therapy, whereas in the non-treated group of 164 patients there was a slight permanent mean weight increase. At the end of 75 months the group not on estrogen therapy was about 3.5 ± 1.8 kg heavier than the group of estrogen-treated women. Such prospective investigations will have to be repeated in a greater number of patients before one can be certain that estrogens are capable of preventing the usual post-climacteric weight increase.

Hirsutism

Estrogens are anti-androgenic and there are also some anti-androgenic progestagens which may reduce the signs of hyperandrogenism, such as hirsutism, de-feminization or even virilism in the post-menopausal woman [BERBLINGER, 1925; PRILL and LAURITZEN, 1970]. Treatment with estrogens in combination with anti-androgens is successful in post-menopausal hirsutism with hyperandrogenism as long as treatment is continued. No investigations have so far been done which show clearly that prophylactic administration of estrogens alone can prevent the development of post-menopausal hirsutism. This seems possible, however, when all available circumstantial evidence is considered.

Long-Term Supervision

An important benefit of every long-term treatment is the close medical supervision under which the patient is kept. This results in the early detection of diseases, in this case particularly of cancer, and may favourably influence morbidity and mortality rates.

References

AERTGEERTS, J.: Influence d'un traitement œstrogène sur la peau de femmes ménopausées ou castrées. Brux. méd. *3:* 215 (1972).

AITKEN, J. M.: Bone metabolism in postmenopausal women; in BEARD The menopause. A guide to current research and practice (MTP Press, Lancaster 1976a).

AITKEN, J. M.: Osteoporosis and its relation to estrogen deficiency; in CAMPBELL The management of the menopause and post-menopausal years, p. 225 (MTP Press, Lancaster 1976b).

AITKEN, J. M.; HART, D.; ANDERSON, J. B.; LINDSAY, R.; SMITH, D. A., and SPEIRS, C. F.: Osteoporosis after oophorectomy for nonmalignant disease in premenopausal women. Br. med. J. *ii:* 325 (1973a).

AITKEN, J. M.; HART, D., and LINDSAY, R.: Oestrogen replacement therapy for the prevention of osteoporosis after oophorectomy. Br. med. J. *iii:* 515 (1973b).

AITKEN, J. M.; HART, D.; LINDSAY, R., and MACDONALD, E. B.: Oestrogen dosage in the treatment of postmenopausal osteoporosis. Medikon int. *8:* 3 (1974).

ALFRAM, P. A.: An epidemiologic study of cervical and trochanteric fractures of the femur in an urban population (Munksgaard, Copenhagen 1964).

ANDERSON, H. E.: Clinical use of estrogen in uterine prolapse. J. Am. med. Ass. *168:* 173 (1958).

ASK-UPMARK, E.: Life and death without ovaries. Acta med. scand. *172:* 129 (1962).

AYLWARD, M.: Plasma tryptophan levels and mental depression in postmenopausal subjects. Effects of oral piperazine-oestrone sulphate. Med. Sci. *1:* 30 (1973).

BEARD, R. S.: Estrogens and the cardiovascular system; in The menopause (MTP Press, Lancaster 1976).

BENEDICT, R. B.: Onset of early course of essential hypertension. J. chron. Dis. *4:* 3 (1956).

BENGTSSON, C.: Ischaemic heart disease in women. Acta med. scand., suppl. 549 (1973).

BERBLINGER, W.: Klimakterische Gesichtsbehaarung und endokrine Drüsen. Z. Konstit. Forsch. *10:* 412 (1925).

BERKSON, D. M.; STAMLER, J., and COHEN, D. B.: Ovarian function and coronary atherosclerosis. Clin. Obstet. Gynec. *7:* 504 (1964).

BIEREN, R. E.: Painful sex in women. Sexology *30:* 88 (1963).

BLANC, J.: Menopause and myocardial infarction. Am. J. Obstet. Gynec. *127:* 353 (1977).

BOL, J. J.: Clinical observations concerning the use of estriol. Ned. Tijdschr. Geneesk. *104:* 1468 (1960).

BOLDING, O. T. and WILLCUTT, H. C.: Physiological and psychological evaluation of the estrogen deprived patient. J. med. Ass. Ala. *39:* 459 (1969).

BORGLIN, N. E.: The effect of estriol on the female genital tract. Acta obstet. gynec. scand. *38:* 157 (1959).

BOYD, G. S.: Oestrogens, cholesterol metabolism and atherosclerosis; in VAN KEEP and LAURITZEN Ageing and estrogens, p. 74 (Karger, Basel 1973).

BROWN, R. K.: Evaluation of dermatologic creams containing female hormones. J. Am. med. Wom. Ass. *21:* 493 (1966).

BROWN, A. D. G.: Postmenopausal urinary problems; in Clinics in obstetrics and gynaecology, vol. 4, p. 181 (Saunders, London 1977).

BUERGER, M.: Geschlecht und Krankheit (Lehmanns, München 1958).

BURCH, J. C.; BYRD, B. F., and VAUGHN, W. K.: The effect of long-term estrogen on hysterectomized women. Am. J. Obstet. Gynec. *118:* 778 (1974).

CAINE, M. and RAZ, S.: The role of female hormones in stress incontinence. Abstr. Int. Soc. Urol., Amsterdam 1973, p. 30.

CALDWELL, B. M.: An evaluation of the psychological effects of sex hormones in aged women. I. Results after six months. J. Geront. *7:* 185 (1952).

CALDWELL, B. M. and WATSON, R.: An evaluation of the psychological effects of sex hormones in aged women. II. Results after eighteen months. J. Geront. *9:* 168 (1954).

CAMPBELL, S.: Double blind psychometric studies on the effects of natural estrogens on post-menopausal women; in CAMPBELL The management of the menopause and post-menopausal years, p. 149 (MTP Press, Lancaster 1976).

CAMPBELL, S. and WHITEHEAD, M.: Oestrogen therapy and the menopause syndrome; in Clinics in obstetrics and gynecology, vol. 4, p. 31 (Saunders, London 1977).

CATTOOR, J. P.; CRABBÉ, G., et ENGELS, J. A.: Utilisation du succinate d'œstriol dans la postménopause. Brux. méd. *4:* 303 (1970).

CHIEFI, M.: An investigation of the effects of parenteral and topical administration of steroids and the elastic properties of senile skin. J. Geront. *5:* 17 (1950).

COCKERELL, E. G. and KNOX, A.: Dermatologic diseases of the vulva (1969). Am. J. Obstet. Gynec. *81:* 537 (1962).

COOPE, J.: Double blind cross-over study of estrogen replacement therapy; in CAMPBELL The management of the menopause and post-menopausal years, p. 159 (MTP Press, Lancaster 1976).

CRANE, M. G. and HARRIS, J. J.: Plasma renin activity and aldosterone excretion rate in normal subjects. J. clin. Endocr. Metab. *29:* 550 (1969).

DAPUNT, O.; FÖDISCH, H. J., und ZEIGER: Lichen sclerosus et atrophicus mit genitaler und extragenitaler Lokalisation. Geburtsh. Frauenheilk. *31:* 661 (1971).

DAVIS, M. E.; JONES, R., and JAROLIM, C.: Long term estrogen substitution and atherosclerosis. Am. J. Obstet. Gynec. *82:* 1003 (1961).

DAVIS, M. E.; LANZL, L. H., and COX, A. B.: Detection, prevention and retardation of postmenopausal osteoporosis. Obstet. Gynec., N. Y. *36:* 187 (1970).

DAXELMUELLER, L.; LAURITZEN, C., und HEILBRUNN, A.: Therapie des klimakterischen Syndroms mit Östriol-Succinat. Z. Therapie *12:* 14 (1974).

DEQUEKER, J.; BURSSENS, A.; CREYTENS, G., and BOUILLON, R.: Ageing of bone: its relation to osteoporosis and osteoarthrosis in postmenopausal women; in VAN KEEP and LAURITZEN Estrogens in the post-menopause, p. 116 (Karger, Basel 1975).

DEQUEKER, J. and FERIN, J.: Effect of long-term estrogen therapy on bone remodelling in women with a natural menopause. Cross-sectional and longitudinal study; in VAN KEEP, GREENBLATT and ALBEAUX-FERNET Consensus on menopause research, p. 124 (MTP Press, Lancaster 1977).

DEWHURST, C. J.: Frequency and severity of menopausal symptoms; in The management of the menopause and post-menopausal years, p. 25 (MTP Press, Lancaster 1976).

DUEKER, H.: Leistungssteigerung durch Keimdrüsenhormone (Barth, München 1957).

EIFF, A. W. VON: Blood pressure and estrogens; in VAN KEEP and LAURITZEN Estrogens in the postmenopause. Front. Hormone Res., vol. 3, p. 177 (Karger, Basel 1975).

EVANS, R. B. and MARMORSTON, J.: Improved mental functioning with premarin therapy in atherosclerosis. Proc. Soc. exp. Biol. Med. *113:* 698 (1963).

FASTNER, Z.: Combined hormonal treatment of urine incontinence in senile women. 10th Int. Congr. Gerontol. Section, Session No. C4 (1975).

FEDOR-FREYBERGH, P.: The influence of oestrogens on the well-being and mental performance in climacteric and postmenopausal women. Acta obstet. gynec. scand., suppl. 64 (1977).

FURMAN, R. H.: Gonadal steroid effect on serum lipids; in SALHANIK, KIPNIS and VAN DE WIELE Metabolic effects of gonadal hormones and contraceptive steroids, p. 247 (Plenum Press, New York 1969).

FURUHJELM, M.: Cited from NORDIN, B. E. C.: Estrogens and bones; in VAN KEEP, GREENBLATT and ALBEAUX-FERNET Consensus on menopause research, p. 49 (MTP Press, Lancaster 1976).

GAD, C.; FISCHER, S., and LEBECH, P.: Changes in the vaginal mucosa during treatment with estradiol and megestrol acetate. Acta obstet. gynec. scand. *50:* 27 (1971).

GALLAGHER, J. C. and NORDIN, B. E. C.: Oestrogens and calcium metabolism; in VAN KEEP and LAURITZEN Ageing and estrogens. Front. Hormone Res., vol. 2, p. 98 (Karger, Basel 1973).

GALLAGHER, J. C. and NORDIN, B. E. C.: Effects of oestrogen and progestogen therapy on calcium metabolism in post-menopausal women; in VAN KEEP and LAURITZEN Estrogens in the post-menopause. Front. Hormone Res., vol. 3, p. 150 (Karger, Basel 1975).

GITSCH, E. und GOLOB, E.: Zur Frage der idealen Östriolkonzentration in Salben bei genitaler Anwendung. Zentbl. Gynäk. *12:* 454 (1962).

Goldzieher, M. A.: The effect of estrogen on the senile skin. J. Geront. *1:* 196 (1946).

Gordan, G. S.: Postmenopausal osteoporosis, cause, prevention and treatment; in Clinics in obstetrics and gynaecology, vol. 4, p. 169 (Saunders, London 1977).

Greenblatt, R. B.: Hormone factors in libido. J. clin. Endocr. Metab. *3:* 305 (1943).

Greenblatt, R. B.: Menopause and its management; in Dorfman and Castro Pituitary-ovarian endocrinology, p. 159 (Holden Day, San Francisco 1963).

Greenblatt, R. B.: Estrogen therapy for postmenopausal females. New Engl. J. Med. *272:* 305 (1965).

Greenblatt, R. B.; Barfield, W. E.; Garner, J. F.; Calk, G. L., and Harrod, J. P.: Evaluation of an estrogen, androgen, estrogen-androgen combination and a placebo in the treatment of the menopause. J. clin. Endocr. *10:* 1547 (1950).

Hällström, T.: Sexuality in the climacteric; in Greenblatt and Studd Clinics in obstetrics and gynaecology, vol. 4, p. 227 (Saunders, London 1977).

Harrison, R. F.: Urethral profile studies on menopausal women and the effects of estrogen treatment; in Campbell The management of the menopause and post-menopausal years, p. 209 (MTP Press, Lancaster 1976).

Hauser, G. A.; Müller, T.; Valaer, M.; Erb, H.; Obiri, J. A.; Remen, U., und Vanäänen, P.: Der Zusammenhang zwischen gynäkologischen Erkrankungen und dem Menopausenalter. Gynaecologia *152:* 270 (1961).

Heaney, R. P.: Pathophysiology of osteoporosis: implications for treatment. Tex. Med. *70:* 37 (1974).

Heaney, R. P. and Recker, R. R.: Estrogen effects on bone remodelling at menopause. Clin. Res. *23:* 535 (1975).

Herrnberg, C. A.: Treatment of postmenopausal osteoporosis with oestrogens and androgens. Acta endocr., Copenh. *34:* 51 (1960).

Higano, N.; Robinson, R. W., and Cohen, W. D.: Increased incidence of cardiovascular disease in castrated women. New Engl. J. Med. *268:* 1123 (1963).

Hoffmann, F.: Konservative Behandlung des Harnröhrenprolaps. Zentbl. Gynäk. *42:* 78 (1950a).

Hoffmann, F.: Über den Einfluss von Östradiolkristallen auf die Funktion der weiblichen Harnblase. Geburtsh. Frauenheilk. *11:* 838 (1950b).

Hunter, D. J. S.; Akande, E. O.; Carr, P., and Stallworthy, J.: The clinical and endocrinological effect of oestradiol implants at the time of hysterectomy and bilateral salpingo-oophorectomy. J. Obstet. Gynaec. Br. Commonw. *80:* 827 (1973).

Jaszmann, L.: Epidemiology of climacteric and postclimacteric complaints; in van Keep and Lauritzen Ageing and estrogens. Front. Hormone Res., vol. 2, p. 22 (Karger, Basel 1973).

Jaszmann, L. J.: Epidemiology of the climacteric syndrome; in The management of the menopause and post-menopausal years, p. 11 (MTP Press, Lancaster 1976).

Jern, H. Z.: Hormone therapy of the menopause and aging. (Thomas, Springfield 1973).

Johansson, B. W.; Kaij, L.; Kullander, S.; Lennér, H.-C.; Svanberg, L., and Astedt, B.: On some late effects of bilateral oophorectomy in the age range 15–30 years. Acta obstet. gynec. scand. *54:* 449 (1975).

JÖNSSON, G.: Treatment of climacteric cystourethritis using long acting oestriol. Klimakteriet Symp. Danish Soc. Obstet. Gynec. (Danmarks Apotekerforening, Helsingoer 1973).

JOSWIG-PRIEWE, H.; JOSWIG, E. H., und OTTO, K.: Orale Östrioltherapie bei Kolpitis senilis. Z. Allgemeinmed. *49:* 633 (1973).

KANNEL, W. B.; HJORTLAND, M. C.; MCNAMARA, R., and GORDON, T.: Menopause and risk of cardiovascular disease. The Framingham study. Ann. intern. Med. *85:* 447 (1976).

KANTOR, H. I.; MICHAEL, G. M.; SHORE, H., and LUDVIGSON, H. W.: Administration of estrogens to older women, a psychosomatic evaluation. Am. J. Obstet. Gynec. *101:* 658 (1968)

KEEP, P. A. VAN and GREGORY, A.: Sexual relations in the ageing female; in MONEY and MUSAPH Handbook of sexology, chap. 62, p. 838 (Excerpta Medica, Amsterdam 1977).

KINSEY, C. C.; POMEROY, W. B.; MARTIN, C. E., and GEBHARD, P. H.: Sexual behaviour in the human female (Saunders, Philadelphia 1953).

KLAIBER, E. L.; BROVERMAN, D. M.; VOGEL, W., and KOBAYASHI, Y.: The use of steroid hormones in depression; in ITIL, LAUDALM and HERRMANN Psychotropic action of hormones, p. 135 (Spectrum, New York 1976).

KNOWELDEN, J.; BUHR, A. J., and DUMBAR, O.: Incidence of fractures in persons over 35 years of age. Br. J. prev. soc. Med. *18:* 130 (1964).

KOPERA, H.: Estrogens and psychic functions; in VAN KEEP and LAURITZEN Ageing and estrogens. Front Hormone Res., vol. 2, p. 118 (Karger, Basel 1973).

KORTE, W.: Indikationen und Praxis vaginaler Oestrogen-Therapie. Fortschr. Med. *91:* 921 (1973).

KROKEFORS, E.: Geburtenzahl und gynäkologische Krankheiten (Liitios, Helsingfors 1960).

KRONIG, O. J. G.: Recent experiences with estriol. Ned. Tijdschr. Geneesk. *103:* 1947 (1959).

KUPPERMAN, H. S.: The menopausal woman and sex hormones. Med. Aspects Human Sex. *1:* 64 (1967).

KUPPERMAN, H. S.; BLATT, M. H. G.; WIESBADEN, H., and FILLER, W.: Comparative clinical evaluation of estrogenic preparations by the menopausal and amenorrhoeal indices. J. clin. Endocr. Metab. *13:* 88 (1953).

LABHART, A.: Die Erkrankungen der äusseren Geschlechtsorgane; in SEITZ und AMREICH Biologie und Pathologie des Weibes, vol. 4, p. 1 (Urban & Schwarzenberg, Berlin 1955).

LACHNIT, U.: Zur Therapie des Prämenopausensyndroms. Medsche Mitt. *31:* 2 (1970).

LAMBILLON, J.: La thérapeutique par l'œstriol dans l'incontinence urinaire à l'effort chez la femme. Brux. méd. *51:* 277 (1971).

LARSSON-COHN, U.: Lipids and estrogens; in VAN KEEP, GREENBLATT and ALBEAUX-FERNET Consensus on menopause research (MTP Press, Lancaster 1976).

LAURITZEN, C.: Erfahrungen mit einem injizierbaren Östriol-Depot-Präparat. Münch. med. Wschr. *19:* 1203 (1968).

LAURITZEN, C.: Das Klimakterium. Physiologie, Pathophysiologie, Diagnose, Differentialdiagnose, Therapie. Tägliche Praxis *11:* 91 (1970).

LAURITZEN, C.: Pathologie des Klimakteriums; in KNÖRR, BELLER und LAURITZEN Lehrbuch der Gynäkologie, p. 117 (Springer, Berlin 1972).

LAURITZEN, C.: The management of the pre-menopausal and the post-menopausal patient; in VAN KEEP and LAURITZEN Ageing and estrogens, p. 2 (Karger, Basel 1973).

LAURITZEN, C. and MUELLER, P.: Pathology and involution of the genitals in the ageing female; in MONEY and MUSAPH Handbook of sexology, chap. 63, p. 846 (Excerpta Medica, Amsterdam 1977).

LAURITZEN, C. and VELIBESE, S.: Clinical investigations concerning long-acting oestriol (polyestriolphosphate). Acta endocr., Copenh. *38:* 72 (1961).

LEBECH, P. E. and BORGGAARD, B.: Changes in the serum lipids and antithrombin III effected by synthetic and natural estrogen therapy; in LEBECH, ULRICH and HELLES Klimakteriet (Danmarks Apotekerforening, Helsingoer 1973).

LIEPELT, A.-O.: Die Symptomatologie des Klimakteriums und ihre Beziehung zur Gesamtmedizin (Enke, Stuttgart 1952).

LINDSAY, R.: Longterm prevention of postmenopausal osteoporosis by oestrogen-evidence for decreased bone mass after delayed onset of oestrogen treatment. Lancet *i:* 1038 (1976).

LONDON, W. T.; ROSENBERG, S. E.; DRAPER, J. W., and ALMY, T. P.: The effect of estrogens on atherosclerosis. A post-mortem study. Ann. intern. Med. *55:* 63 (1961).

MANCHESTER, J. H.; HERMAN, M. V., and GORLIN, R.: Premenopausal castration and documented coronary atherosclerosis. Am. J. Cardiol. *28:* 33 (1971).

MARKS, R. and SHAHRAD, P.: Skin changes at the time of the climacteric; in GREENBLATT and STUDD Clinics in obstetrics and gynecology, vol. 4, p. 207 (Saunders, London 1977).

MASTERS, W. H. and JOHNSON, V. E.: Human sexual response (Little, Brown, Boston 1966).

MCEVEN, D. C.: Ovarian failure and the menopause. Can. med. Ass. J. *92:* 962 (1965).

MCKINLAY, S. M. and JEFFERIES, M.: The menopausal syndrome. Br. J. prevent. soc. Med. *28:* 108 (1974).

MCLAREN, H. C.: The normal menopause. J. Obstet. Gynaec. Br. Commonw. *48:* 1 (1941).

MEEMA, H. E.; BUNKER, H., and MEEMA, S.: Preventive effect of estrogen on postmenopausal bone loss. Archs intern. Med. *135:* 1436 (1975).

MICHAEL, C. M.; KANTOR, H. J., and SHORE, H.: Further psychometric evaluation of older women. The effect of estrogen administration. J. Geront. *25:* 337 (1970).

MONEY, J.: Sex hormones and other variables in human eroticism; in YOUNG Sex and internal secretions (Williams & Wilkins, Baltimore 1961).

MONEY, J. and DALÉRY, J.: Sexual disorders: hormonal and drug therapy; in MONEY and MUSAPH Handbook of sexology, chap. 103, p. 1303 (Excerpta Medica, Amsterdam 1977).

MOORE, B.; GUSTAFSON, R., and STUDD, J.: Experience of a National Health Service Menopause Clinic. Curr. Med. Res. Opin. *3:* suppl. 3, p. 42 (1975).

NEUGARTEN, B. L. and KRAINES, R. J.: Menopausal symptoms in women of various ages. Psychosom. Med. *27:* 266 (1966).

NORDIN, B. E. C.: Report on workshop: Estrogens and bones; in VAN KEEP, GREENBLATT and ALBEAUX-FERNET Consensus on menopause research, p. 48 (MTP Press, Lancaster 1976).

NORDIN, B. E. C.; GALLAGHER, J. C.; AARON, J. E., and HORSMAN, A.: Postmenopausal osteopenia and osteoporosis; in VAN KEEP and LAURITZEN Estrogens in the postmenopause. Front. Hormone Res., vol. 3, p. 131 (Karger, Basel 1975).

NOTELOWITZ, M.: The effect of natural estrogens on blood pressure and weight in postmenopausal women. S. Afr. med. J. *49:* 2251 (1975).

NOTELOWITZ, M.: The effect of long term oestrogen replacement therapy on glucose and lipid metabolism in postmenopausal women. S. Afr. med. J. *50:* 2001 (1976).

NOVAK, E.: The menopause. J. Am. med. Ass. *156:* 575 (1954).

NOVAK, E. and WILLIAMS, T. J.: Autopsy comparison of cardiovascular changes in castrated and normal women. Am. J. Obstet. Gynec. *80:* 863 (1960).

OLIVER, M. F.: Ischaemic heart disease in young women. Br. med. J. *iv:* 253 (1974).

OLIVER, M. F. and BOYD, G. S.: Effect of bilateral ovariectomy on coronary-artery disease and serum lipid levels. Lancet *ii:* 690 (1959).

PARRISH, H. M.; CARR, C. A.; HALL, D. G., and KING, T. M.: Time interval from castration in premenopausal women to development of excessive coronary atherosclerosis. Am. J. Obstet. Gynec. *99:* 155 (1967).

PFEFFER, R. J. and VAN DEN NORT: Estrogen use and stroke risk in postmenopausal women. Am. J. Epidem. *103:* 445 (1976).

PFEIFFER, E. and DAVIS, G. C.: Determinants of sexual behavior in middle and old age. J. Am. Geriat. Soc. *20:* 151 (1972).

PICKERING, G. W.: High blood pressure (London 1955).

PINCUS, G.: The control of fertility (Academic Press, New York 1965).

PITKIN, R. M. and BRADBURY, J. T.: The effect of topical estrogen on irradiated vaginal epithelium. Am. J. Obstet. Gynec. *92:* 175 (1965).

POCKRANDT, H.: Wirkungen eines synthetischen Östrogens (Dienöstrol-diazetat) auf die Haut bei perkutaner Verabreichung. Z. Geburtsh. Gynäk. *141:* 85 (1954).

POTOCKI, J.: The effects of oestrogen treatment on coronary insufficiency in women after the menopause. Polski Tygod. lek. *26:* 1812 (1971).

PRIJOT, E.; BARZIN, L., et DESTEXHE, B.: Essai de traitement hormonal de la kératoconjonctivité sèche. Bull. Soc. belge Ophtal. *162:* 795 (1972).

PRILL, H. J.: Blutdruck und Klimakterium. Geburtsh. Frauenheilk. *26:* 338 (1966).

PRILL, H. J. und LAURITZEN, C.: Das Klimakterium; in SCHWALM und DÖDERLEIN Klinik der Frauenheilkunde und Geburtshilfe, vol. 8, p. 339 (Urban & Schwarzenberg, München 1970).

PROOST, J.; DEMOL, R., and BARBIER, F.: Oestriolbehandeling van de menopauze en de postmenopauze. Bull. Soc. belge Gynéc. Obstét. *39:* 65 (1969).

PUNNONEN, R. and RAURAMO, L.: The effect of castration and oral estrogen therapy on serum lipids; in VAN KEEP, GREENBLATT and ALBEAUX-FERNET Consensus on menopause research, p. 132 (MTP Press, Lancaster 1976).

RAURAMO, L.: in VAN KEEP and LAURITZEN Estrogens in the post-menopause, Discussion, p. 182 (Karger, Basel 1975).

RAURAMO, L.: Non-genital target tissues of estrogens; in VAN KEEP, GREENBLATT and ALBEAUX-FERNET Consensus on menopause research, p. 32 (MTP Press, Lancaster 1976).

Rauramo, L.; Lagerspetz, K.; Engblom, P., and Punnonen, R.: The effect of castration and peroral estrogen therapy on some psychological functions; in Estrogens in the post-menopause. Front. Hormone Res., vol. 3, p. 94 (Karger, Basel 1975).

Rauramo, L. and Punnonen, R.: The effect of castration and peroral estrogen therapy on a woman's skin; in van Keep and Lauritzen Ageing and estrogens, p. 48 (Karger, Basel 1973).

Riggs, B. L.; Jowsey, J.; Goldsmith, R. S.; Kelley, P. J.; Hoffman, D. L., and Arnaud, C. D.: Short and long-term effects of oestrogen and synthetic anabolic hormone in postmenopausal osteoporosis. J. clin. Invest. *51:* 1659 (1972).

Ritterband, A. B.; Jaffe, J. A.; Densen, P. M.; Magagna, J. F., and Reed, E.: Gonadal function and development of coronary heart disease. Circulation *27:* 237 (1963).

Rivin, A. O. and Dimitroff, S. P.: Incidence and severity of atherosclerosis in estrogen treated males and females with hypooestrogenic state. Circulation *9:* 533 (1954).

Rosenberg, L.; Armstrong, B., and Jick, H.: Myocardial infarction and estrogen in postmenopausal women. New Engl. J. Med. *294:* 1256 (1976).

Royal College of General Practitioners: Oral contraceptives and health (Pitman, London 1974).

Ruette, B. von und Delnon, J.: Die Urethritis atrophicans der Frau. Helv. chir. Acta *32:* 484 (1965).

Salmon, U. J.; Walter, R. E., and Geist, S. H.: The use of oestrogens in the treatment of dysuria and incontinence in postmenopausal women. Am. J. Obstet. Gynec. *42:* 845 (1941).

Shahrad, P. and Marks, R.: The effect of estrogens in the skin; in Campbell The management of the menopause and post-menopausal years, p. 243 (MTP Press, Lancaster 1976).

Sheffery, J. B.; Wilson, T. A., and Walsh, J. C.: Double blind, cross-over study comparing chlordiazepoxide, conjugated estrogens, combined chlordiazepoxide and conjugated estrogens, and placebo in treatment of the menopause, p. 443. Med. Ann. Distr. Columbia *38:* 433 (1969).

Simon, M. J.: Cystalgies de la postménopause et œstriol. Presse méd. *69:* 1301 (1961).

Slunsky, R.: Das Ovestin als konservative Behandlungsmethode der funktionellen Harninkontinenz. Z. Geburtsh. Gynäk. *160:* 91 (1963).

Smith, D. A.; Anderson, J. B.; Shimmins, J.; Speirs, C. F., and Barnett, E.: Changes in metacarpal mineral content and density in normal male and female subjects with age. Clin. Radiol. *20:* 23 (1969).

Smith, P. J. B.: The effect of estrogens on bladder function in the female; in Campbell The management of the menopause and post-menopausal years, p. 201 (MTP Press, Lancaster 1976).

Smith, P. J. B.: Menopause and lower urinary tract – another case for hormonal replacement therapy? Practitioner *218:* 97 (1977).

Spellacy, W. N. and Birk, S. A.: The effect of intrauterine devices, oral contraceptives, estrogens, and progestagens on blood pressure. Am. J. Obstet. Gynec. *112:* 912 (1972).

Spitzer, R. S.; Kyu-Taik, L., and Thomas, W. A.: Early age of menopause in young women with total acute myocardial infarction. Am. Heart J. *53:* 805 (1957).

Studd, J.; Chakravarti, S., and Oran, D.: The climacteric; in Clinics in obstetrics and gynaecology, vol. 4, p. 3 (Saunders, London 1977).

Strohfeld, P.: Klinische Untersuchungen über Stoffwechselwirkungen von Östriol-succinat. Drug Res. *27:* 2006 (1977).

Sznajderman, M. and Oliver, M. F.: Spontaneous premature menopause, ischaemic heart-disease and serum lipids. Lancet *i:* 962 (1963).

Utian, W. H.: The mental tonic effect of oestrogens administered to oophorectomized females. S. Afr. med. J. *46:* 1079 (1972a).

Utian, W. H.: Effects of oophorectomy and estrogen therapy on serum cholesterol. Int. J. Gynaec. Obstet. *10:* 95 (1972b).

Utian, W. H.: Definitive symptoms of post-menopause – incorporating use of vaginal parabasal cell index; in van Keep and Lauritzen Estrogens in the post-menopause. Front. Hormone Res., vol. 3, p. 74 (Karger, Basel 1975).

Utian, W. H.: The scientific basis for post-menopausal estrogen therapy. The management of specific symptoms and rationale for long term replacement; in Beard The menopause, p. 175 (MTP Press, Lancaster 1976)

Utian, W. H.: Current status of menopause and postmenopausal estrogen therapy. Obstetl. gynec. Surv. *32:* 4 (1977).

Vanhulle, G. and Demol, R.: A double blind study into the influence of estriol on a number of psychological tests in post-menopausal women; in van Keep, Greenblatt and Albeaux-Fernet Consensus on menopause research, p. 94 (MTP Press, Lancaster 1976).

Wallach, S. and Hennemann, P. H.: Prolonged estrogen therapy in postmenopausal women. J. Am. med. Ass. *171:* 1637 (1959).

Wallner, O. und Soost, H. J.: Klinische und vaginal-zytologische Untersuchungen über die Wirkung von Östriol-Succinat. Med. Klin. *42:* 1413 (1971).

Walter, S.; Walf, H.; Barlebo, H., and Jensen, K.: Urinary incontinence in post-menopausal women treated with oestrogens. Abstr. Int. Continence Soc. Antwerp 1976.

Wied, G. L.: Über die Bedeutung der Suggestion in der Therapie von klimakterischen Ausfallsbeschwerden. Ärztl. Wschr. *8:* 623 (1953).

Wuest, J.; Dry, T., and Edwards, J.: The degree of coronary atherosclerosis in bilateral oophorectomized women. Circulation *7:* 801 (1953).

Dr. C. Lauritzen, Department of Gynaecology and Obstetrics. University of Ulm, Prittwitzstrasse 43, *D-7900 Ulm/Donau* (FRG)

Discussion

HASPELS: In your review of the beneficial effects of estrogens you spoke in a general way about the various estrogens available. Could you elaborate on the different estrogens, particularly estriol?

LAURITZEN: This could be the subject of a separate lecture, but let me say here that every treatment, certainly one for climacteric complaints, should be individually designed for each patient. In my experience the therapeutic effects of conjugated estrogens are not greatly different from those of, for example, estradiol valerate or micronized estradiol. Further, there is a tendency, of which I do not approve, simply to double the dose of conjugated estrogens to 2.5 mg/day when an initial dose of 1.25 mg fails to alleviate all a patient's complaints. In some cases it may well be that an increase of 20 % would have been sufficient and that the increase of 100 % will lead to the occurrence of secondary effects. I do think that estriol has a special place. In therapeutically effective dosages it does not stimulate the endometrium nor does it increase the triglyceride levels.

UTIAN: There is more to therapy than the alleviation of complaints. There is also the mental-tonic effect of estrogens. There does seem to be a dose-dependent response here; we have seen an effect with doses of 1.25 mg or more of conjugated estrogens, but this has disappeared when the dose was lowered to 0.625 mg.

BAULIEU: Prof. LAURITZEN's paper was on the effects of estrogens, and not so much on replacement therapy to substitute ovarian endocrine function. The estrogens used are not the same as produced by the ovary in the fertile years and the mode of administration is by no means identical to the cyclical ovarian activity. Also the effects of therapy on other changes in the endocrine pattern after the menopause have not been considered: effects on adrenal function, pituitary function, etc. A real replacement therapy should be aimed at restoring all the functions changed by the menopause.

LAURITZEN: I have spoken about estrogen therapy and not about substitution, because it is, at least in case of the post-menopausal woman, not possible to substitute ovarian function completely and precisely. Substitution may, however, be desirable and even necessary in the case of young castrated women and certain risk patients. I shall speak about these later. I do not mean, however, that when administering estrogens one should disregard the basic course of the ovarian cycle; one should certainly administer estrogens cyclically and, if it is feasible, add progesterone.

NORDIN: Prof. LAURITZEN said that it was difficult to quantify the symptomatology of the climacteric phase. At Leeds, however, we have made an attempt to do this. We made a list of 18 questions concerning symptomatology ranging from hot flushes and sweats, via the nervous complaints to questions concerning libido, orgasm, dryness of the vagina and dyspareunia. For each question a score could be given from zero to three, for nothing, slight, moderate, and severe, respectively. The minimum score therefore was zero and the total maximum 54. It was found that young women, those between the ages of 20 and 39, scored an average of less than 10, the score rose after the age of 40, and was highest in the group of women of menopausal age, particularly amongst those attending the menopause clinic.

Analysis showed the most important discriminant between pre- and post-menopausal women to be hot flushes. The second most important discriminant are the sexual complaints, which were not included in the Blatt Menopausal Index. The third most important are muscle pains, aching joints and back pain. The fourth are the feelings of depression, of loneliness, and lack of confidence. Dryness of the skin and facial hair have no discriminant value.

This system of analysis has also proved to be of value in the assessment of the effect of therapy. We have found, for example, that on estrogen therapy half the patients recovered their libido in 6 months.

LAURITZEN: BLATT gave different weights to different complaints. You have done this?

NORDIN: No, we have not done this yet, but our discriminant function analysis shows what weightings could be applied to the various symptoms if we wish to develop the questionnaire in that direction.

Front. Hormone Res., vol. 5, pp. 26–39 (Karger, Basel 1978)

Application of Cost-Effectiveness Analysis to Post-Menopausal Estrogen Therapy

W. H. Utian

Department of Reproductive Biology, Case Western Reserve University School of Medicine, Cleveland, Ohio

Introduction

During a decade of personal research into menopause and the role of estrogen replacement therapy the pendulum of popular opinion has swung from full right to full left and back again. In a 1968 review of the potential risks and benefits, contrary to popular opinion, I remarked on the lack of confirmative evidence for both, and stated that: 'most of what is published is based on emotional and philosophical premises; the change of life is an emotional subject not only to women, but to men and doctors'. I warned that: 'Ideally these potent hormones should not be administered by medical practitioners in the absence of definite climacteric symptoms until there is clearer evidence of the relationship between the development of carcinoma and the administration of estrogens' [17].

Recently a strong trend against estrogen therapy has developed. Paradoxically I find myself yet again in conflict with popular opinion. This does not imply that I refute newly reported findings, but simply that I question their interpretation within the overall context of the problem.

In ten years a previously neglected area has become a subject of intensive research. Every new answer, perhaps inevitably, has posed yet more questions, so that debate and discussion have become even more acrimonious now than in the past. It is therefore time for us to put the subject in perspective; for a balanced evaluation of specific benefits and actual risks to be made.

An attempt at evaluation of the risk-benefit ratio of the long-term effects of pure estrogens does highlight the fact that, compared to the great many carefully designed large scale epidemiologic studies on the health consequences of oral contraceptive use, no comparable body of data is available for pure estrogen. Despite this lack, it is possible to subject these risks and benefits, and their potential costs, to some form of mathematical evaluation and analysis, and to come out with some answer or direction.

My present purpose, therefore, is to deal with the principles and application of general analytic methods to the subject of long-term estrogen therapy after the menopause. Although I find myself delicately treading a minefield of conflict, emotion and lack of data, I do believe a general analytic approach is mandatory if we are to make headway of any sort. At the very least such analysis can act as a tool to define areas of deficient data, and at best it may restore equanimity to an overcharged debate.

Some Comments on General Analytic Approaches

Various measures have been designed that can help guide present and future decision making by systematic analysis [5, 24]. For example, answers may be needed to problems in allocation of limited health care resources or as aids to health care decision makers in planning new facilities. As mentioned, the balancing of benefits of estrogen therapy against potential risks is a subject in need of such analysis. The answer we seek in this instance should help physicians decide whether such therapy should be given or not.

To be of value, any form of analysis must be comprehensive and broadly applicable. This necessitates the availability of the best current information on both the efficacy of therapy and its costs, as well as the possible risks and the respective costs thereof. Unfortunately the available data base on the effectiveness of most clinical procedures, estrogen therapy and menopause included, is distressingly limited. This is not reason to despair. Nor is the tendency among physicians to demand exact scientific proof, although this is commendable and desirable. In point of fact, until such objectives are obtained, current analysis and decision making must depend upon the best available current evidence. Nonetheless, any form of analysis should be structured to incorporate new data as becomes available and even to suggest areas in need of future research to resolve critical uncertainties [24].

An ideal measure of effectiveness of a clinical practice needs to be outcome orientated, with length and quality of life as the ultimate measures. Inevitably where risks and benefits of a particular form of therapy are being evaluated, and in turn involve possible trade-offs between longevity and quality of life, then subjective values have to be involved.

Another area that needs incorporation into any effective analysis is the balance between present and future health benefits and costs. Estrogen therapy, for example, can be considered as a preventative medicine program in which the costs are immediate and ongoing, but the health benefits and risks may be in the future.

Specific Analytic Approaches

Cost-benefit analysis [5] and cost-effectiveness analysis [24] are two different analytic approaches to assessment of health practices.

Cost-benefit analysis. Values all outcomes in economic (e. g. dollar) terms. This requires that human lives and quality of life be valued in dollars. Once the benefits and costs have been reduced to dollar values, the decision whether to administer a form of therapy, in theory at least, is simplified.

Cost-effectiveness analysis. Priority is placed on alternate expenditures without requiring that the dollar value of life and health be assessed [24]. *Health effectiveness* is measured in 'quality-adjusted life years' or QALYs because it incorporates changes in survival and morbidity in a single measure that reflects trade-offs between them. This will be further explained below. *Health care costs* are measured in dollars. The measure of *cost-effectiveness* then becomes a ratio, the numerator being cost in dollars and the denominator being net health effectiveness measured in QALYs. Thus the lower the ratio, the better the result.

Net Health Care Costs:

There are various ways to measure the cost of medical care:

Direct (therapeutic) costs. Includes payment for service by doctors, hospitals, laboratories, pharmacy, etc.

Indirect costs. This is a more comprehensive measure that takes into account the impact of illness, premature death or disability on the economy. Being broader in concept it is less easy to define.

Preventive costs. Costs of a prophylaxis program which includes drugs, dispensing, education, physician time, etc. Savings on such a program can be calculated by deducting the actual preventive costs from the hypothetical costs of treating the disease in the absence of a prophylactic program.

This difference could also be termed *avoidable costs.*

The following formula for calculation of health costs, modified from WEINSTEIN and STASON [24] takes such factors, but not indirect costs, into account:

$$\triangle C = \triangle C_{RX} + \triangle C_{SE} - \triangle C_{BENEF} + \triangle C_{RX\triangle LE},$$

where $\triangle C$ = net health care costs (HCC); $\triangle C_{RX}$ = all direct HCC (drugs, physician, etc.); $\triangle C_{SE}$ = all HCC due to side effects of treatment; $\triangle C_{BENEF}$ = savings in HCC due to disease prevention; $\triangle C_{RX\triangle LE}$ = HCC of diseases that would not have occurred if patient did not receive treatment.

Net Health Effectiveness

The measure of improvement or loss in quality of life is a controversial area. A weighting scheme or Health Status Index (λS) [6,24] assigns a weight (P) of zero to one for varying degrees of disability or discomfort to full health. The greater the number the worse the disability, i. e. 0.0 is virtually no disability and 1.0 would imply that to the patient the condition is about as bad as being dead, i. e. the probability (P) for death is 1.0. The health status index (λS) is then calculated as $1-P$.

The following examples demonstrate the use of the Health Status Index to quantify the situation for a potential risk like uterine cancer:

Morbidity	P
A = Early cancer removed; no disability	0.05
B = Surgery plus radiotherapy; no disability	0.15
C = Vaginal stenosis and pain	0.35
D = Severe pain, recurrence	0.88
E = Secondary disease	1.0
$\lambda S = 1-P$	
and $\lambda S \times$ years = $\triangle Y_{CANCER}$ *in* QALY	

and for a potential benefit like prevention of osteoporosis:

Morbidity	P
A = Radiologic osteoporosis; no disability	0.1
B = Occasional discomfort on exertion	0.35
C = Fractured wrist	0.55
D = Femoral neck or vertebral compression fracture	0.85
E = Totally bedridden, disabled, intractable pain	1.0
$\lambda S = 1-P$	
and $\lambda S \times$ years = $\triangle Y_{OSTEO}$ *in* QALY	

The number of years spent (YS) at this health status is multiplied by λS to yield the number of *quality adjusted life years* (QALYs). This can be considered to be equivalent to the number of years spent in full health. The calculation is thus as follows:

$$\lambda S \times YS = \triangle Y_{BENEF} \text{ or } \triangle Y_{SE} \text{ in QALY},$$

where λS = HSI or disability weight = 1−P if P = probability for death; YS = years at health status; $\triangle Y_{BENEF}$ = net health effectiveness; $\triangle Y_{SE}$ = net health loss; QALY = quality-adjusted life years.

The overall measure of improvement or loss in quality of life, or Net Health Effectiveness, can be calculated from the following formula, also modified from WEINSTEIN and STASON [24]:

$$\triangle E = \triangle Y + \triangle Y_{BENEF} - \triangle Y_{SE},$$

where $\triangle E$ = net health effectiveness in QALY; $\triangle Y$ = expected no. unadjusted life years; $\triangle Y_{BENEF}$ = improvement in quality of life years due to reduction in morbidity or prevention thereof; $\triangle Y_{SE}$ = loss due to side effects of treatment.

The expected number of unadjusted life years ($\triangle Y$) can be derived from life tables. The measure of $\triangle Y_{BENEF}$ and $\triangle Y_{SE}$ are calculated as previously described.

The data is not usually available to make these measures with certainty, a source of criticism of cost-effectiveness analysis. The whole trade-off concept is difficult. Nonetheless, such trade-offs in therapeutic measures are made daily by physicians. All that health analysis does is supply a quantification, that is to make the measure explicit.

Cost-Effectiveness Analysis Applied to Estrogen Therapy After the Menopause

The direct value therefore of cost-effectiveness analysis is that it forces the physician or health care administrator to be explicit in evaluating risks and benefits as they apply to any therapeutic modality. Let us now attempt to apply this concept to long-term estrogen therapy after the menopause.

Calculation of the equation is dependent upon numerous local factors in varying countries and communities and therefore needs independent analysis. Indeed much of the data to be presented later in the present workshop will influence the answers. I will therefore do no more than define the areas of risk and benefit and make some remarks on cost.

$\triangle Y_{BENEF}$ – *Health Benefits – Alleviation or Prevention of Morbidity*

There is a fair degree of evidence for the efficacy of long-term estrogen therapy, reviewed in depth elsewhere [23]. Unfortunately, estimates of these benefits in terms of mortality and morbidity probabilities are sorely lacking. It is therefore difficult to calculate with confidence the differences in quality-adjusted life expectancy. This problem becomes more obvious as we examine each of the potential benefits.

Relief of Symptoms

Relief of true estrogen deficiency symptoms [18] is limited to hot flushes and those related to vaginal atrophy. The measure of a Health Status Index and calculation of QALYs for these two parameters is subjective and difficult and there is no solid data on which QALY can be calculated. Nonetheless, relief of symptoms significantly influences quality of life and time at work, and must be considered in cost-effectiveness analysis.

Prevention of Osteoporosis

The prevention of osteoporosis and its complications is one of the most convincing indications for long-term estrogen therapy, and the subject for in-depth discussion elsewhere in this workshop. In order to calculate health effectiveness it is necessary to have good stastistics for incidence rates of this problem in treated and not treated groups.

Well-defined prevalence, complication and disability rates are hard to come by. Nonetheless, the problem is an enormous one. For example, Gallanaugh *et al.* [10] reported an incidence of 1/1,000 males (514 cases) and 3.17/1,000 females (2,105 cases) affected by femoral neck fractures per year, a fact necessitating the equivalent of a 250-bed hospital to cope with the situation for that area. The average length of stay in hospital was 34.2 days; the long-term disability was not described. The data presented by Stewart [16] and Bauer [2] is similar.

The in-depth analysis of the relative frequency and problems of osteoporosis, to be presented by Prof. Nordin at this workshop, will provide a good foundation for cost-effectiveness analysis in this area.

A disability weight or Health Status Index for osteoporosis has been presented earlier but needs modification by further discussion.

Prevention of Atheromatosis and Coronary Heart Disease

Prevention of coronary heart disease no longer appears to be a justifiable indication for long-term estrogen therapy [3, 20]. Nor, however, can it be considered to be a direct cause of the problem [15]. CHD should therefore be excluded from the 'trade-offs' in assessment of cost-effectiveness analysis.

Prevention of Ageing

In this context 'ageing' is a nebulous term that needs definition and quantification. Until this can be achieved, specific benefits, for example, change in skin texture, need be considered and analyzed. Overall analysis of ageing *per se,* therefore has no role in this particular cost-effectiveness analysis. Much research is still necessary in this area.

The Mental Tonic Effect

The mental tonic effect of long-term estrogen therapy described by UTIAN [19] in 1972 has been confirmed [9]. This is an important component of health effectiveness with its direct bearing on quality of life and the ability of the individual to function as an effective member of society and of the total available labor force. Some 50% at least of women show this positive response [19] and the contribution of this effect to $\triangle Y_{BENEF}$ in a cost-effectiveness analysis is considerable.

$\triangle Y_{SE}$ – *Treatment Risks and Side Effects*

It is worthy of reemphasis that cost-effectiveness analysis takes present knowledge as a data base, and allows future knowledge to be incorporated. Potential risks must be evaluated with this in mind. Thus the question is not, for example, whether estrogen causes cancer, but rather, within present knowledge, what is the worst possible risk that may be entailed if estrogen does cause cancer. As further evidence becomes available, so the $\triangle Y_{SE}$ component will be recalculated and readjusted to the new lower or higher value.

There is one major problem in risk analysis. It is extremely difficult to assess the risk attributable to one factor like estrogen without considering the synergistic effects of other health risk factors such as hypertension, smoking and obesity and possible others which are not easily identified. For example, contraceptive pill users who are nonsmokers are at very much lower risk of heart disease mortality than women who use pills and smoke cigarettes [13]. Does the same apply to pure estrogen? Moreover, these synergistic influences vary in different countries and within different socioeconomic groups of the same countries. It is not always obvious in many studies whether such possible synergistic effects have been considered. Certainly, policymakers involved in evaluating the risk-benefit ratio of long-term estrogen use are well advised to determine the risk that applies to use in their own country, and should use caution in extrapolating directly from studies done elsewhere.

Potential risks of long-term estrogens have been reviewed in depth elsewhere [23]. Those to be included in any cost-effectiveness analysis are as follows:

Postmenopausal bleeding. This is a real risk of postmenopausal estrogen therapy. The incidence, however, will vary dependent upon prescribing practices, particularly the drug selection (e. g. estriol vs. estradiol) and dosage.

Unnecessary surgery. The increased incidence of postmenopausal bleeding may directly affect the incidence of diagnostic curettage and possibly hysterectomy. For instance, it is known that the rate of hysterectomies within the United States increased from 602/100,000 in 1970 to 727/100,000 in 1975 [8]. What is not known is the influence, if any, of estrogen therapy on this increase. It is crucial that such procedures be considered in the analysis for their effect on health and health costs.

Uterine cancer. Here the pertinent problem, certainly in terms of current knowledge, is to define the maximal extent of the potential risk. The annual incidence of endometrial malignancy for all ages in the United States is 0.21/1,000 women [14]. 1/1,000 untreated postmenopausal women with an intact uterus will develop endometrial cancer a year [25]. Several studies have indicated an increased risk for estrogen users amounting to as many as 8 cases per 1,000 estrogen users per year (see review by GREENWALD *et al.* [11]). At worst, therefore, the risk appears to be an extra 7 cases per 1,000 per annum and this figure should be the current one to be incorporated in the analysis of cost-effectiveness.

Breast cancer. The data base for an increased incidence of breast cancer is unsound. Nonetheless in terms of present knowledge the relative risk is increased by a factor of 1.3 at 12 years estrogen usage, and of 2 at 20 years usage [12]. These figures, therefore, should be incorporated in the measure of $\triangle Y_{SE}$.

Deep vein thrombosis and thromboembolism. Unlike oral contraceptive usage, there are no good statistics for deep venous thrombosis and thromboembolism on pure estrogen. The risk does appear to be doubled in postmenopausal estrogen users as compared to nonusers, i. e. from 9/100,000 to 18/100,000 [4].

Gallbladder disease. The incidence rate of gallbladder disease requiring surgical treatment is increased 2.5-fold in women taking estrogen after menopause. The Boston Collaborative Drug Surveillance Program [4] put these figures at 87/100,000 per year in healthy women aged 45–49 not on estrogens and 218/100,000 for estrogen users. The net difference to be incorporated into the cost-effectiveness analysis would appear therefore to be 131/100,000 estrogen users per year. This is likely to have a considerable impact.

The Risk-Benefit Ratio

Despite the absence of the necessary arithmetic, it is obvious that the trade-off between risks and benefits, or risk:benefit ratio, is finely balanced. The logical answer is to apply what I prefer to term the Minimax Concept – aim to minimize risks and maximize benefits.

The application of the Minimax Concept goes beyond the bounds of the present paper, but is worthy of introduction, because much can be done to reduce the potential risks without impairing potential benefits, even in terms of present knowledge. For example, different estrogens appear to induce different metabolic responses [20–22]. Selective drug usage such as avoidance of ethinyl estradiol because of increased thrombogenic potential, or greater use of estriol because of decreased endometrial stimulation, may help reducc the therapeutic risk of thromboembolism and uterine cancer respectively.

As these areas become clarified, so the cost-effectiveness analysis will need updating.

Health Care Costs

In a world beset with problems of inflation, disparity of health programs and variable health needs, it would be more than presumptuous to attempt to advise on the desirability of a general estrogen prophylaxis program and to estimate costs for such a program for any specific region. It is mandatory, of course, that costs be calculated as per the formula already presented, but that local statistics be utilized.

Some limited attempts have been made at such analysis. For example, AITKEN [1] and DEWHURST [7] have independently estimated that the total annual drug bill for Britain would be about £ 90,000,000 if conjugated estrogens were routinely prescribed for prophylaxis. DEWHURST [7] has calculated the bill for medical services would add a further £ 65,000,000 per year. AITKEN [1] in turn calculated that the annual cost of treating femoral neck and Colles' fractures in a stable population of 200,000 people (in which population there would be about 38,000 women over 45) would be about £ 40,000 or £ 10,000,000 for Britain per year. This would allow about £ 1 per head per annum before the cost of prophylaxis exceeded the likely cost of treating complications of osteoporosis as they arose. Unfortunately, the British figures do not take into account the Health Status Index, that is the cost to the patient of being in pain or disabled or unable to live an independent existence. Recalculation of the figures with true cost-effectiveness analysis would therefore be of greater value. In Britain, nonetheless, the cost of prevention of osteoporosis clearly far exceeds the cost of treatment of the developed problem.

GREENWALD *et al.* [11] have estimated that 14 % of American women over the age of 45 now take estrogens, predominantly conjugated estrogen (Premarin) at a cost of $ 82,777,000, in 1975 shipments for use in the United States. AITKEN [1] estimated that use of Premarin in Britain could cost about £ 88,500,000 whereas cheap synthetics would cost about £ 2,600,000 and stated that 'clearly until such time as Premarin is proven to be superior to ethinyl estradiol the prescribing of this expensive extract of horse urine should be discouraged'.

It is apparent that controversy exists. Different estrogens do appear to exert different effects [20, 21]. The answer therefore, in addition to usage of local statistics, is an urgent need for cost-effectiveness analysis to be applied to the use of all the presently available types of estrogen as well.

Conclusions

Application of cost-effectiveness analysis to long-term estrogen therapy should be of value in that it will force physicians and health planners to be explicit about the beliefs and values that underlie their decisions. Where points of view differ, the relative trade-offs can be compared more directly.

There is an urgent need for better data on the overall efficacy, risks and costs of therapy with different estrogens. Nonetheless, application of presently available data is possible and necessary.

Future clinical trials and epidemiological observation should take cost-effectiveness analysis into account.

At the present time the decision to prescribe long-term estrogen therapy must rest upon a choice between relying on a proper analysis despite the present imperfections, or on no analysis at all. To quote WEINSTEIN and STASON [24]: 'The former, in these times of increasingly complex decisions, difficult tradeoffs, and limited resources, is by far the preferred choice.'

References

1 AITKEN, J. M.: Bone metabolism in post-menopausal women; in BEARD The menopause, pp. 95–142 (MTP, Lancaster 1976).

2 BAUER, G. C. H.: Epidemiology of fracture in aged persons. Clin. Orthop. *17:* 219–225 (1960).

3 BLANC, J.; BOSCHAT, J.; MORIN, J.; CLAVIER, J., and PENTHER, P.: Menopause and myocardial infarction. Am. J. Obstet. Gynec. *127:* 353–355 (1977).

4 Boston Collaborative Drug Surveillance Program: Surgically confirmed gall-bladder disease, venous thromboembolism and breast tumours in relation to postmenopausal estrogen therapy. New Engl. J. Med. *290:* 15–19 (1974).

5 BUNKER, J. P.; BARNES, B. A., and MOSTELLER, F.: Costs, risks and benefits of surgery, pp. 5–108 (Oxford University Press, New York 1977).

6 BUSH, J. W.; CHEN, M. M., and PATRICK, D. L.: Health status index in cost-effectiveness analysis of PKU program; in BERG Health Status Indexes: Proceedings of a conference conducted by Health Services Research, Tucson, Arizona (Hospital Research and Educational Trust, Chicago 1973).

7 DEWHURST, C. J.: The role of estrogen in preventive medicine; in BEARD The menopause, pp. 219–245 (MTP, Lancaster 1976).

8 Editorial: Rate of hysterectomy increases. Ob. Gyn. News *12:* 2 (1977).

9 FURUHJELM, M. and FEDOR-FREYBERGH, P.: The influence of estrogens on the psyche in climacteric and postmenopausal women; in VAN KEEP, GREENBLATT, ALBEAUX-FERNET The menopause, pp. 84–93 (MTP, Lancaster 1976).

10 GALLANAUGH, S. C.; MARTIN, A., and MILLARD, P. H.: Regional survey of femoral neck fractures. Br. med. J. *ii:* 1496–1497 (1976).

11 GREENWALD, P.; CAPUTO, T. A., and WOLFGANG, P. E.: Endometrial cancer after menopausal use of estrogens. Obstet. Gynec., N. Y. *50:* 239–243 (1977).

12 HOOVER, R.; GRAY, L. A.; COLE, P., and MACMAHON, B.: Menopausal estrogens and breast cancer. New Engl. J. Med. *294:* 401–405 (1976).

13 JAIN, A. K.: Mortality risk associated with the use of oral contraceptives. Stud. Fam. Plann. *8:* 50–54 (1977).

14 LEVIN, D. L.; DEVESSA, S. S.; GODWIN, J. D., and SILVERMAN, D. T.: Cancer rates and risks, p. 13 (Department of Health, Education and Welfare, Washington 1974).
15 ROSENBERG, L.; ARMSTRONG, B., and JICK, H.: Myocardial infarction and estrogen therapy in post-menopausal women. New Engl. J. Med. *294:* 1256–1259 (1976).
16 STEWART, I. M.: Fractures of neck of femur: incidence and implications. Br. med. J. *i:* 698–701 (1955).
17 UTIAN, W. H.: Feminine forever? Current concepts on the menopause. A critical review. S. Afr. J. Obstet. Gynaec. *6:* 7–10 (1968).
18 UTIAN, W. H.: The true clinical features of post-menopause and oophorectomy and their response to oestrogen therapy. S. Afr. med. J. *46:* 732–737 (1972).
19 UTIAN, W. H.: The mental tonic effect of oestrogens administered to oophorectomized females. S. Afr. med. J. *46:* 1079–1082 (1972).
20 UTIAN, W. H.: Effects of oophorectomy and oestrogen therapy on serum cholesterol. Int. J. Gynec. Obstet. *10:* 95–101 (1972).
21 UTIAN, W. H.: Effects of oophorectomy and subsequent oestrogen therapy on plasma calcium and phosphorus. S. Afr. J. Obstet. Gynaec. *10:* 8–16 (1972).
22 UTIAN, W. H.: Comparative trial of P1496, a new non-steroidal oestrogen analogue. Br. med. J. *i:* 579–581 (1973).
23 UTIAN, W. H.: Current status of menopause and postmenopausal estrogen therapy. Obstet. Gynec. Surv. *32:* 193–204 (1977).
24 WEINSTEIN, M. C. and STASON, W. B.: Foundations of cost-effectiveness analysis for health and medical practices. New Engl. J. Med. *296:* 716–721 (1977).
25 WEISS, N. S.: Risks and benefits of estrogen use. New Engl. J. Med. *293:* 1200 (1975).

Dr. W. H. UTIAN, Assistant Professor, Department of Reproductive Biology, Case Western Reserve University School of Medicine, *Cleveland, OH 44106* (USA)

Discussion

LAURITZEN: Dr. UTIAN mentioned the increased likelihood of hospitalization for curettage as a factor affecting the cost:effectiveness ratio. In our clinic we have found that women receiving estrogen therapy are 4.7 times more likely than others to be hospitalized for curettage and that they are three times more likely than others to be hysterectomized. It may be that this is to some extent the result of the therapy, though it is also likely that it is because these women are under the regular surveillance of a gynaecologist, whereas others are not.

UTIAN: Such figures may vary slightly from country to country depending on local health practices, but it is important to know them and to take them properly into account when calculating the cost of the therapy.

LAURITZEN: Certainly, when contemplating operating or performing a curettage, the fact that a woman is or is not receiving estrogens is an important consideration.

UTIAN: This is the point that I am trying to emphasize: the patient on estrogen is definitely at an increased risk of undergoing some form of surgery. This increases her health risks in terms of the surgery she undergoes, and it increases the costs. Balanced against this is the fact that such surgery may prevent some future complications for that particular patient, so there will be a health benefit as a result of those costs and risks. This all comes into the equation.

VAN KEEP: Is it not likely that the woman who will eventually receive estrogens in her post-menopause already has a different pattern in her illness behaviour from the woman who will not? The incidence of hysterectomy is not evenly distributed over the various social class groups and neither is the use of estrogens.

UTIAN: This is true. Doctors' wives, for instance, in the United States have a higher incidence of hysterectomy than the average population. The same is likely to be so of patients who see their physicians frequently. The indication for the type of surgery also varies from one country to another. In the United States the indication is often a questionable endometrial interpretation, perhaps based on the fears of litigation. If the patient has an atypical hyperplasia, it may be interpreted as a carcinoma or potential carcinoma. These are all factors that, country by country, have to be taken into account.

MCKAY HART: It has been said that estrogen therapy can bring about no improvement in the case of osteoporosis. In the experience of our group in Glasgow, however, a small dose of mestranol, 20–40 μg/day, can in fact result in some improvement in the bone density even where bone has already been lost. The difference has been found to be statistically significant even where the treatment has been begun 6 years after oophorectomy. This benefit is one which should be taken into account when calculating the cost:effectiveness of the therapy.

NORDIN: Dr. UTIAN said that it is difficult to quantitate the relief of symptoms and I wonder if the questionnaire I mentioned after the presentation of the previous paper could not be useful in this connection. Secondly, I am not convinced that the cost of the estrogens used in this therapy is a major factor in these equations. Certainly the amount spent on these, at any rate in Great Britain, is absolutely trivial when compared to that spent on, for instance, tranquillizers.

UTIAN: It is important to have an index. It is important, however, to separate the true estrogens deficient components from coincidental ageing factors. If an index like yours could be made more precise and could separate the psycho-socio-cultural, coincidental ageing factors from true hormone deficiency factors, then it would indeed be very useful. As to your second point regarding the cost of the estrogen preparations themselves, I cannot help but feel that such costs, relatively small though they may be, are becoming more and more important. In the past there was a tendency for us to think that the world had unlimited resources and that the resources available for health care were also unlimited. It is becoming increasingly clear, however, that this is not so, and gradually all countries are realizing the need to set a limit on the amount of money available for health matters. As this happens, priorities have to be allocated, and it stands to reason that the projects with viable cost:effectiveness ratios will be the ones most likely to receive financial support. It is then that all costs, even relatively minor ones, become significant.

NORDIN: As far as the index is concerned, the features which respond to estrogen therapy are, by definition, the ones that are relevant to this discussion.

UTIAN: We have made a very broad evaluation of the true symptoms of estrogen deficiency. Around 30 possible symptoms were assessed in different groups: oophorectomized and non-oophorectomized, pre-menopausal and post-menopausal, and, in the various post-menopausal groups, on and off different therapies and on placebo. The only *specific* estrogen deficient complaints found were hot flushes and related symptoms, and vaginal atrophy. All the others: anxieties, depression, and loss of libido, and so on, were all related to age or to social or cultural factors; they were not true estrogen deficiency symptoms.

COMNINOS: What exactly do you mean by 'long-term' estrogen therapy when you speak of it in connection with your evaluation of the cost:benefit ratio?

UTIAN: The expression 'long-term' is open to interpretation, but when I use it here I do so to refer to a therapy to be used for life from the onset of menopausal problems. This is another point which should be considered when one is working on the cost:effectiveness equation.

Front. Hormone Res., vol. 5, pp. 40–52 (Karger, Basel 1978)

Factors Relating to Morbidity and Mortality of Women on Long-Term Estrogen Therapy

A. E. Rakoff

Jefferson Medical College, Thomas Jefferson University, Philadelphia, Pa.

As is the case with all potent therapeutic agents, the administration of estrogen is associated with the occurrence of side effects and hazards. Until recently the most common reasons for discontinuance of estrogen replacement therapy by menopausal and postmenopausal women were the occurrence of side effects such as weight gain, fluid retention, mastalgia and nausea, whereas now large numbers of patients are discontinuing or refusing to start estrogen therapy because of the fear of relatively uncommon but more serious potential hazards. In part these risks have been garnered from studies on women taking oral contraceptives and in part from retrospective studies on postmenopausal women on estrogen replacement therapy. Differences attributable to the dose and duration of estrogen usage, the progestin component in the oral contraceptives, the age group of the patients as well as many other variables undoubtedly introduce confounding effects, which require consideration in evaluating the estrogen risk. The present status of those hazards which will be summarized in this presentation include hypertension, thromboembolic phenomena, myocardial infarction, gallbladder disease, endometrial carcinoma and breast lesions.

Hypertension (table I)

Hypertension has been cited as one of the potential hazards of estrogen therapy. However, this contention has been based chiefly on findings in oral contraceptive users in whom a number of investigators have found slightly higher mean blood pressure readings than in nonusers, a finding which is reversible on discontinuation of the pills. In their study, Fisch

Table I. Hypertension

Investigators	Medication	Increased risk
Fisch and Frank	oral contraceptives	3×
Royal College Study	oral contraceptives	2.5× [1]
Pfeffer and Van den Noort	estrogen replacement	2.37×
Stern et al.	estrogen replacement	none

[1] Correlated with progestin dosage, duration of use, age, but not with estrogen dosage.

and FRANK [10] reported that the age-adjusted proportion of oral contraceptive users with a blood pressure over 140/90 mm Hg was about three times that of nonusers.

PFEFFER and VAN DEN NOORT [22] reported a risk ratio of 2.37 for hypertension in postmenopausal estrogen users in an evaluation of 1,034 women in a retirement community. There was, however, no significant increase in risk for stroke. On the other hand, in a study by STERN *et al.* [27], in which women taking only estrogen were compared with oral contraceptive users and nonusers, no significant increase in systolic or diastolic blood pressure was observed in estrogen-taking and nonestrogen-taking women, although an increase in both systolic and diastolic pressures were also noted by them in oral contraceptive users. Possible explanations for the observed difference in these groups may be attributed either to the amount of estrogen taken, which is generally less for estrogen replacement therapy, or to the synthetic progestagens present in oral contraceptives. In the oral contraceptive study of the Royal College of General Practitioners [23] the incidence of hypertension appeared to correlate with the amount of progestagen in the pills and not with the amount of estrogen; there was also a positive correlation with the duration of use and age.

Myocardial Infarction (table II)

It was previously hoped that the administration of estrogen to postmenopausal women might decrease their risk of developing atherosclerotic cardiovascular disease. The presumption for this protective effect was based on the well known sex difference which had been traditionally attributed to higher estrogen levels in premenopausal women, and to observations

Table II. Myocardial infarction

Investigators	Medication	Increased risk
Mann et al.	oral contraceptives	3×[1]
Royal College Study	oral contraceptives	uncertain
Boston Collaborative Study	estrogen replacement	none
Pfeffer and Van den Noort	estrogen replacement	none

[1] About 2.8 from 20–39 and 4.7 from 40–44 years.

suggesting that early menopause or castration increased the risk of myocardial infarction. However, doubt concerning such a protective effect and the possibility that the administration of exogenous estrogen may even increase the risk has now been raised. Statistics from the Coronary Drug Project [7] indicated that the administration of relatively large doses of conjugated equine estrogens (5 mg/day) increased the risk of nonfatal, recurrent myocardial infarction in men. The study by MANN *et al.* [19] pointed to an increased risk of about threefold of fatal and nonfatal myocardial infarction in women taking oral contraceptives. Although the risk estimate was similar to women in the younger age group, total mortality attributable to complications was considerably greater among women over the age of 40.

In a report from the Boston Collaborative Surveillance Program [3, 4] data were analyzed to correlate the relationship between estrogen use and nonfatal myocardial infarction in postmenopausal women. The data were obtained from 24 Boston-area hospitals for 1972 and also from 21 hospitals in a number of different countries for patients admitted since 1969. Postmenopausal women between 40 and 75 years of age were included in the analyses, except for those women who had used estrogens irregularly. A total of 336 myocardial infarction patients were compared with a reference group of 6,730 patients. 8 of the 336 myocardial infarction patients (2.4%) and 330 reference patients (4.9%) were regular estrogen users, for a crude rate ratio of 0.47 at the time of hospitalization. After control for confounding variables, including age, past history of myocardial infarction, angina, diabetes, hypertension and cigarette smoking, the summary point estimate rate ratio was 0.97, with 95% confidence limits of 0.48 and 1.95. It was concluded from these data that there was no evidence of a statistically significant association between current regular use of estrogens and nonfatal acute myocardial infarction.

No evidence is known of either a beneficial or adverse effect of estrogens on the probability that postmenopausal women will survive a myocardial infarction.

Thromboembolic Disease (table III)

An increased incidence of venous thrombosis, including deep vein thrombosis in the legs and cerebrovascular disease in women who used oral contraceptives has been reported in both retrospective and prospective studies [3, 23]. In the retrospective study on 1,310 women by the Collaborative Group for Study of Stroke in Young Women [6] the findings indicated a fourfold increase in the risk of thrombotic stroke. The incidence of venous thrombosis has been found to be lower with preparations containing 50 μg of ethinyl estradiol or mestranol compared with those containing higher doses [15].

In contrast to these findings in oral contraceptive users no significant association between estrogen use by postmenopausal women and idiopathic venous thrombosis was found in the Boston Collaborative Drug Surveillance Program [4]. A probable explanation for this difference is the estrogen dose, since this has been found to be a significant factor in the oral contraceptives [15]. Most postmenopausal estrogen users take a smaller dose than that present in the low dose oral contraceptives.

In the Boston Collaborative study the overall incidence rate of idiopathic venous thrombosis was estimated to be 11/100,000/year. It was concluded that the general belief that the risk of venous thromboembolism increases with age is probably a function of a higher incidence of predis-

Table III. Thromboembolic disease

Investigators	Medication	Increased risk
Royal College Study	oral contraceptives	5× (venous thrombosis) 4× (cerebrovascular)
Collaborative Stroke Study	oral contraceptives	4× (stroke)
Boston Collaborative Study	estrogen replacement	none (venous thrombosis)

posing diseases. From the present data suggesting a dose-related relationship of estrogen use and thromboembolism it would appear judicious to use the smallest effective dose.

Gallbladder Disease (table IV)

An association between the use of oral contraceptives and cholecystitis and cholangitis is now generally accepted. It has also been claimed that gallstones occur twice as often in patients who use oral contraceptives, and that the incidence is related to the duration of use [23]. The rate appears to rise after 2 years of pill usage and plateaus after 4–5 years.

A similar increase in gallbladder disease has been reported in postmenopausal women on estrogen therapy by several investigators [4]. In the report from the Boston Collaborative Drug Surveillance Program the relative risk was found to be 2.5 times higher in postmenopausal women who used estrogens than in those who did not. Factors such as age, parity, obesity or prior hysterectomy did not appear to explain the association. There was no evidence of a relationship with duration of use, although in users of oral contraceptives the risk appeared to be highest after 6–12 months of use. The overall risk is small, estimated to be about 79 per 100,000 users per year [23].

Of interest with regard to the influence of ovarian hormones on the formation of gallstones are the two recent studies by BENNION and coworkers [1, 2]. The first showed that cholesterol saturation of bile was increased with the use of oral contraceptives and the second showed a change in bile lipids accompanying oophorectomy in a premenopausal woman, in which the molar percentage of cholesterol decreased after oophorectomy. These findings suggest a biochemical basis for the increase in gallbladder disease among women using oral contraceptives and estrogens.

Table IV. Gallbladder disease

Investigators	Medication	Increased risk
Royal College Study	oral contraceptives	2× (gallstones)
Boston Collaborative Study	estrogen replacement	2.5× (gallstones)

Endometrial Cancer (table V)

At the present time the chief concern regarding the use of estrogens for menopausal and postmenopausal women is that of an increased risk for the development of endometrial cancer based on a series of five recently reported retrospective studies [13, 17, 18, 25, 29]. The findings are in contrast to the frequently quoted 1967 report by DUNN and BRADBURY [8] in which no differences were noted in frequency of estrogen use as a factor in the occurrence of endometrial carcinoma based on a study of patients admitted for uterine bleeding. Although the validity of the recently reported retrospective studies indicating a possible causal association of estrogen use and endometrial carcinoma has been questioned on a number of scores [11] most editorial comments in prestigious journals have intimated that the current epidemiologic evidence is sufficiently suspicious of a causal relationship, as to warrant present increased surveillance, as well as further study. In addition the US Food and Drug Administration has recently modified the labelling of estrogens to require full warning of the risk of endometrial cancer for each estrogen prescription.

Among the objections which have been raised to the recent studies are: That they were not double-blind and allocation to case or control groups were not randomized; that they were retrospective studies, and therefore, subject to the criticism of association versus causal significance; that the groups were unrepresentative of the female population; that other variables may contribute to the association; that preexisting lesions may have been present in the hormone-treated group; greater frequency of examinations among the estrogen-treated group; misinterpretation or difference of opinion of the pathologic diagnosis. In addition, STECKEL [26] has raised the important point that the preexisting diagnosis that required the use of estrogens, which could be the most important discriminant fac-

Table V. Endometrial carcinoma

Investigators	Medication	Risk ratio
Smith et al.	all estrogens	4.5
Ziel and Finkle	conjugated estrogens	7.6 (5.6–13.9)
Mack et al.	all estrogens	8.0
	conjugated estrogens	5.6
Gray et al.	conjugated estrogens	3.1 (0–11.5)
McDonald et al.	conjugated estrogens	4.9–7.9

tor between patients and controls, was apparently not considered. Obviously additional studies are required to evaluate these factors and some are indeed now under way. In the September 15 (1977) issue of the *New England Journal of Medicine,* GORDON *et al.* [12] reported on the review of the histologic material from the ZIEL and FINKLE [29] study by three internationally recognized pathologists. At least one of the pathologists concurred with the original diagnosis in all but 1 of the 89 cases and in 66 or 74% of the cases the diagnosis of endometrial carcinoma by the three experts was unanimous. On the basis of the 66 patients and 132 matched controls the revised risk-ratio estimate was 8.1, validating the original estimate of 7.6. It is also to be pointed out that in the study by GRAY *et al.* [13] only cases were included in which a diagnosis of invasive carcinoma was corroborated by two well-recognized gynecologic pathologists.

To more definitively answer most of the other objections which have been raised, prospective studies with strict controls are required. One such study is that of the Chelsea Hospital group [28]. It is as yet too early to evaluate the findings with regard to endometrial carcinoma. A disturbing preliminary observation is that in some cases atypical hyperplasia can arise without symptoms with a low cyclical estrogen dosage. As is well known, with higher doses irregular breakthrough bleeding may occur, thus increasing the frequency of diagnostic curettage.

An association of duration and dosage of estrogen use and the incidence of endometrial carcinoma was noted in several of the retrospective studies. In the ZIEL and FINKLE study [29] the risk ratio estimate increased with duration of exposure from 5.6 for 1 to 4.9 years to 13.9 for 7 or more years. In the study by MACK *et al.* [17] a dose-response effect was also noted. At smaller doses cyclic administration appeared to decrease the risk. In the study by MCDONALD *et al.* [18] the estimated relative risk associated with conjugated estrogen treatment of 6 months or longer was 4.9 and this increased to 7.9 with exposure for 3 years or longer. The risk increased in those using larger doses (1.25 mg or more). Of particular interest is the fact that less than 6 months exposure did not increase the risk. Also of interest in this study is that the increased relative risk was less apparent in patients with physiologic characteristics often associated with endometrial carcinoma such as obesity and hypertension.

The question of differences among various estrogens in their effect on the endometrium has been repeatedly raised. This has been suggested on the basis that estrone is the predominant estrogen produced by post-menopausal women, while estradiol is the predominant estrogen in premenopausal women [16]. In its review of this subject the Medical Letter of May, 1976 [20] concludes that the consensus of informed opinion is that no convincing evidence has demonstrated that any one type of synthetic or natural estrogen is less likely to be carcinogenic than any other. This, however, is a point still open to controversy.

Breast Lesions (table VI)

Most epidemiologic studies have failed to show an increased incidence of breast cancer in estrogen users, and indeed, some have shown a lower than expected rate. These include studies on women on oral contraceptives and, also, postmenopausal women on estrogen replacement therapy. In the Boston Collaborative Drug Surveillance study of 1974 [4] it was particularly encouraging to note that there was no suggestion of a higher risk even in long-term postmenopausal estrogen users.

In a recent much publicized study, Hoover *et al.* [14] presented data from a retrospective study which they claim clearly indicates that menopausal estrogen use does not protect against breast cancer and possibly points to a slightly increased risk. Their study was made in 1,891 women given conjugated estrogens for the menopause who were followed for a mean period of 12 years. Overall, 49 cases were observed whereas 39.1 were expected on the basis of rates expected from the general population, giving a relative risk of 1.3. The risk increased with follow-up duration and in women using higher dose tablets, and, peculiarly, in those taking the medication on an other than daily basis. There was also a greater risk among women in whom benign disease developed after they had started the drug. This study has come under criticism because of lack of an adequate control series. In this regard the claim has been made that while the oral contraceptives may decrease the risk of development of benign breast

Table VI. Breast cancer

Investigators	Medication	Increased risk
Royal College Study	oral contraceptives	none
Boston Collaborative Study	oral contraceptives	none
Hoover et al.	estrogen replacement	1.3×

disease, they may increase the incidence of breast cancer in women with already established benign breast disease [21, 24].

Summary and Conclusions

A number of serious hazards have been attributed to long-term estrogen therapy on the basis of clinical evaluations in women taking oral contraceptives or estrogen replacement therapy for the menopause.

There is an increased risk of mild hypertension of 2.5- to 3-fold in women on oral contraceptives which is reversible on withdrawal of the medication. It appears that this may be correlated with progestin dosage rather than with estrogen dosage. In menopausal women on estrogen replacement there is a questionable risk ranging from none to 2.3.

Oral contraceptives appear to be associated with risk ratio up to 3-fold for myocardial infarction. No increase has been noted in association with estrogen therapy for the menopause.

Thromboembolic disease carries a 4–5 risk ratio for thromboembolic phenomena including cerebrovascular disease, while no increased risk has been noted with menopausal estrogen replacement.

There is a 2- to 2.5-fold increase in gallstones in association with both oral contraceptives and long-term estrogen replacement therapy.

Recently, five retrospective studies claimed an association of long-term estrogen therapy with an increased risk ratio for endometrial carcinoma, ranging from a mean of 3 to 8. The association appears to be related to dose and duration.

There is no association of breast cancer and the use of oral contraceptives and a doubtful increase in risk was noted with estrogen replacement therapy in one retrospective study.

In view of the estrogen dose relationship noted in the association with most of the suspected hazards of estrogen replacement therapy prospective studies are urgently required not only for confirmation but to determine whether there is an effective dose and duration which is relatively free of serious hazards.

In this presentation we have discussed hazards for which an increased risk has been attributed to estrogen use. We have not considered hazards whose risks may be lowered by estrogen use [5]. Critical statistical studies of the latter are urgently required in evaluating the overall morbidity and mortality resulting from the use of estrogens.

References

1 Bennion, L. J.: Changes in bile lipids accompanying oophorectomy in a premenopausal woman. New Engl. J. Med. *297:* 709 (1977).

2 Bennion, L. J.; Ginsberg, R. L.; Barnick, M. B., and Bennett, P. H.: Effects of oral contraceptives on the gallbladder bile of normal women. New Engl. J. Med. *294:* 189–192 (1976).

3 Boston Collaborative Drug Surveillance Program: Oral contraceptives and venous thromboembolic disease, surgically confirmed gallbladder disease, and breast tumors. Lancet *i:* 1399–1404 (1973).

4 Boston Collaborative Drug Surveillance Program, Boston University Medical College: Surgically confirmed gallbladder disease, venous thromboembolism, and breast tumors in relation to postmenopausal estrogen therapy. New Engl. J. Med. *290:* 15–18 (1974).

5 Byrd, B. F., jr.; Burch, J. C., and Vaughn, W. K.: The impact of long term estrogen support after hysterectomy: a report of 1,016 cases. Ann. Surg. *185:* 574–580 (1977).

6 Collaborative Group for the Study of Stroke in Young Women: Oral contraception and increased risk of cerebral ischemia or thrombosis. New Engl. J. Med. *288:* 871–878 (1973).

7 Coronary Drug Project Research Group: The coronary drug project: initial findings leading to modifications of its research protocol. J. Am. med. Ass. *214:* 1303–1313 (1970).

8 Dunn, L. J. and Bradbury, J. T.: Endocrine factors in endometrial carcinoma. Am. J. Obstet. Gynec. *97:* 465–471 (1967).

9 Fasal, E. and Paffenbarger, R. S., jr.: Oral contraceptives as related to cancer and benign lesions of the breast. J. natn. Cancer Inst. *55:* 767–773 (1975).

10 Fisch, I. R. and Frank, J.: Oral contraceptives and blood pressure. J. Am. med. Ass. *237:* 2499–2503 (1977).

11 Gordan, G. S. and Greenberg, B. G.: Exogenous estrogens and endometrial cancer. Postgrad. Med. *59:* 66 (1976).

12 Gordon, J.; Reagan, J. W.; Finkle, W. D., and Ziel, H. K.: Estrogen and endometrial carcinoma. New Engl. J. Med. *297:* 570–571 (1977).

13 Gray, L. A.; Christopherson, W. M., and Hoover, R. N.: Estrogens and endometrial carcinoma. Obstet. Gynec. *49:* 385–392 (1977).

14 Hoover, R.; Gray, L. A., sr.; Cole, P., *et al.:* Menopausal estrogens and breast cancer. New Engl. J. Med. *295:* 401–405 (1976).

15 Inmann, W. H. W.; Vessey, M. P.; Westerholm, B., *et al.:* Thromboembolic disease and the steroidal content of oral contraceptives: a report of the committee on safety of drugs. Br. med. J. *ii:* 203–209 (1970).

16 MacDonald, P. C. and Siiteri, P. K.: The relationship between the extraglandular production of estrone and the occurrence of endometrial hyperplasia. Gynecol. Oncol. *2:* 259–263 (1974).

17 Mack, E. M.; Pike, M. C.; Henderson, B. E., *et al.:* Estrogens and endometrial cancer in a retirement community. New Engl. J. Med. *294:* 1262–1267 (1976).

18 McDonald, T. W.; Annegers, J. F.; O'Fallon, W. M.; Dockerty, M. B.; Mal-

KASIAN, G. D., and KURLAND, L. T.: Exogenous estrogen and endometrial carcinoma: case-control and incidence study. Am. J. Obstet. Gynec. *127:* 572–580 (1977).
19 MANN, J. I.; INMANN, W. H. W., and THOROGOOD, M.: Oral contraceptive use in older women and fatal myocardial infarction. Br. med. J. *ii:* 445–447 (1976).
20 The Medical Letter: The new estrogens: are they safe? Vol. 18, No. II, p. 45 (1976).
21 ORY, H.; COLE, P.; MACMAHON, B., *et al.:* Oral contraceptives and reduced risk of benign breast diseases. New Engl. J. Med. *294:* 419–422 (1976).
22 PFEFFER, R. I. and VAN DEN NOORT, S.: Estrogen use and stroke in postmenopausal women. Am. J. Epidem. *103:* 445–456 (1976).
23 Royal College of General Practitioners: Oral contraceptives and health: an interum report from the oral contraceptive study of the Royal College of General Practitioners (Pitman, London 1974).
24 SARTWELL, P. E.; ARTHES, F. G., and TONASCIA, J. A.: Epidemiology of benign breast lesions: lack of association with oral contraceptive use. New Engl. J. Med. *288:* 551–554 (1973).
25 SMITH, D. C.; PRENTICE, R.; THOMPSON, D. J., and HERRMANN, W. L.: Association of exogenous estrogen and endometrial carcinoma. New Engl. J. Med. *293:* 1164–1170 (1975).
26 STECKEL, R. J.: Letter to the editor. New Engl. J. Med. *294:* 848 (1976).
27 STERN, M. P.; BROWN, B. W., jr.; HASKELL, W. L.; FARQUHAR, J. W.; WEHRLE, C. L., and WOOD, P. D. S.: Cardiovascular risk and use of estrogens or estrogen-progestagen combinations. J. Am. med. Ass. *235:* 811–815 (1976).
28 WHITEHEAD, M. I.; MCQUEEN, J.; BEARD, R. J.; MINARDI, J., and CAMPBELL, S.: Acta obstet. gynec. scand. (in press, 1977).
29 ZIEL, H. K. and FINKLE, W. D.: Increased risk of endometrial carcinoma among users of conjugated estrogens. New Engl. J. Med. *293:* 1167–1170 (1975).

Prof. A. E. RAKOFF, Department of Obstetrics and Gynecology, Jefferson Medical College of Thomas Jefferson University, *Philadelphia, PA 19107* (USA)

Discussion

STUDD: In a recently published study into the mortality rate among oral contraceptive users [Royal College of General Practitioners: Lancet *1977:* 8041] it was striking to note that only one of the 56 women who died was a non-smoker. It may well be important to keep this in mind when assessing the risks of estrogen therapy in the post-menopause. In another new study, that of WHITEHEAD, mentioned by Professor RAKOFF, five carcinomas were seen in a group of 127 women, but here two things must be said: firstly, the 127 women in this study can by no means be regarded as being representative of post-menopausal women as many were pre-menopausal attending a 'menopausal clinic' for various problems, and secondly, three of those who had a carcinoma had irregular bleedings before therapy began, and the other two had hyperplasia prior to therapy. I must say too that when a vague and doubtful diagnosis suggesting endometrial carcinoma is made in a woman taking estrogens, it is essential to discontinue the estrogen therapy and to give a progestagen for 14 days. This may change the picture completely.

RAKOFF: This shows the superiority of a prospective study over a retrospective one. In any retrospective study these 5 cases would have been included in the statistics as having been associated with the use of estrogens.

KOPERA: We all accept that prospective studies are superior to retrospective ones, but we must not underestimate the difficulties in conducting prospective trials. This should be remembered particularly if one wants to demonstrate the influence of any therapy on the incidence of effects that only develop after long-term treatment.

Although a prospective study is to be preferred, its performance can become impossible due to excessive costs, extreme duration and/or insurmountable difficulties.

VAN KEEP: DE KAY, who organized the British Royal College of General Practitioners' prospective study on the effects of oral contraceptives recently spoke about the logistical problems of a similar study on estrogens in the post-menopause. He calculated that in order to investigate all possible influences of the therapy it would be necessary to study 50,000 women-years on treatment. In view of the low number of estrogen users in Britain at the present time, such a study is not yet feasible.

RAKOFF: It is clear that we often have to be satisfied with retrospective studies, but these must be properly controlled and the findings handled with special care. Only when enough of them have been done – and done well – as is the case with the studies on the relation between smoking and the incidence of lung cancer, should much credence be attached to them.

LAURITZEN: You mentioned, only in passing, the study of BURCH and BYRD, of which one report appeared in the proceedings of our earlier conference [BURCH, J. C.; BYRD, B. R., and VAUGHN, W. K.: The effects of long-term estrogen administration to women following hysterectomy; in VAN KEEP and LAURITZEN Estrogens in the post-menopause. Front. Hormone Res., vol. 3, pp. 208–214 (Karger, Basel 1975)], but this study is interesting because it showed the effect of estrogen therapy on a number of conditions, such as myocardial infarction, cerebrovascular accidents, and on the incidence of all cancers.

RAKOFF: Interesting though this study is, the problem is that there was no control group. BURCH and BYRD compared the observed cases with statistics for the population as a whole, statistics which may have been valid for the population from which their patients came. Moreover, morbidity statistics are not always very accurate. I agree, however, that this study is important and reassuring.

LEMON: Data from some of the state tumour registers in the United States, those of Oregon and California, for instance, show that, on the west coast at least, endometrial cancer is now more prevalent than cervical cancer, and that the incidence of the former has risen considerably.

RAKOFF: Until 1970 the National Cancer Survey of the United States failed to show any increase in the incidence of endometrial cancer. Since then, however, there have certainly been local reports that the figure has gone up, and it will be interesting from the next National Survey if this is so on a national level. It seems likely.

LAURITZEN: Such an increase should not necessarily be explained by the increased use of estrogen though. ROSOL has recently published figures showing a rise of 65 % in the incidence of endometrial cancer in recent years in Czechoslovakia, a country in which estrogens are hardly used [Zentbl. Gynäk. *98:* 175 (1976)].

KELLER: Personally, I am not happy with all FDA regulations concerning precautions and warnings about the risks of therapy. They could pressurize a physician into doing an hysterectomy before even starting estrogen replacement therapy simply in order to avoid, for instance, the somewhat improbable occurrence of uterine complications. An hysterectomy *is,* after all, major surgery.

RAKOFF: I fully agree. In the United States things are even worse. Many doctors are scared to prescribe anything at all for fear of being sued for substantial amounts should anything untoward, even relatively minor, occur.

STUDD: If the incidence of endometrial cancer has increased, what has happened to the mortality? I thought that – nationwide – it went down.

VOOIJS: That is perhaps not so. It seems that mortality due to endometrial cancer has gone up in the United States, because of an increase in the incidence of a type of cancer that was less frequent before: the adeno-squamous type. The average age at which the usual type of adenomatous cancer occurs is 57–59 but this type occurs at a later age (65) and the mortality is higher.

LAURITZEN: Clinical practices in the United States have tended to be somewhat different from those in other countries. The dosages used in the United States, certainly in the past, were higher, estrogens were given continuously, and, when given in dosages that certainly lead to endometrial proliferation, were given without the periodic addition of progestagens. One last point: patients on estrogens do tend to have a higher drug intake than many other people, and it would appear from some retrospective studies that an increased risk of endometrial cancer may be associated with this as such.

Front. Hormone Res., vol. 5, pp. 53–75 (Karger, Basel 1978)

Post-Menopausal and Corticosteroid-Induced Osteoporosis

R. G. CRILLY, A. HORSMAN, D. H. MARSHALL and B. E. C. NORDIN

Medical Research Council Mineral Metabolism Unit, The General Infirmary, Leeds

Introduction

In previous papers [GALLAGHER and NORDIN, 1973, 1975; NORDIN *et al.*, 1975] we have presented evidence that bone resorption increases at the menopause, that this is due to increased sensitivity of the bone to parathyroid hormone in the absence of oestrogen, that bone is lost from the skeleton as a result and that the process is reversible with oestrogen therapy. In the present paper we present recent additional findings, with particular emphasis on the similarities between post-menopausal and corticosteroid-induced osteoporosis and the role of bone formation.

Methodology

Skinfold thickness was measured with Harpenden callipers three times on the dorsum of each hand over the 2nd, 3rd and 4th metacarpals, and the mean value used. Maturation value was determined on urinary sediment smears by the method of CASTILLO *et al.* [1948]. Plasma androstenedione, oestrone and testosterone were measured by radioimmunoassay on ether extracts of plasma samples obtained at about 9.0 a. m. from fasting subjects. Bone mineralisation rates were determined after intravenous administration of radiocalcium by the method of BURKINSHAW *et al.* [1969]. Bone resorption is the difference between the mineralisation rate and calcium balance.

The subjects studied were classified as normal or osteoporotic by means of vertebral measurements on lateral X-rays of the thoracic and lumbar spine as described by HORSMAN [1976]. Two points were scored for a compressed vertebra and one for a wedged vertebra; only cases with 'scores' of 5 and over were termed osteoporotic. The corticosteroid-treated cases were all taking prednisolone in doses of 5–30 mg daily for asthma, rheumatoid arthritis, polyarthritis, systemic lupus erythematosus or Sjögren's disease. In this paper, no attempt has been made to distinguish between steroid-treated cases with and without osteoporosis.

All the other methods used in this paper have been described elsewhere [NORDIN *et al.*, 1976].

Observations

Epidemiology

A diagrammatic representation of bone loss from different parts of the skeleton in post-menopausal women is shown in figure 1. This process starts at or about the time of the menopause and proceeds at different rates in different bones. Since the normal skeleton comprises about 20% trabecular bone and 80% of cortical bone, and the average individual loses

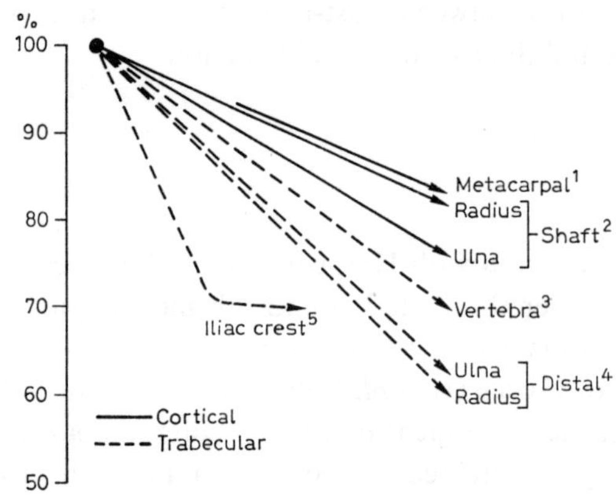

Fig. 1. Diagrammatic representation of bone loss from different parts of the skeleton in post-menopausal women. Note the accelerated loss of trabecular bone. 1 = CA cm^2, 2nd metacarpal midshaft; 2 = CW cm, 4 cm site; 3 = g cm^{-3}; 4 = g cm^{-1}, 2 cm site; 5 = % volume.

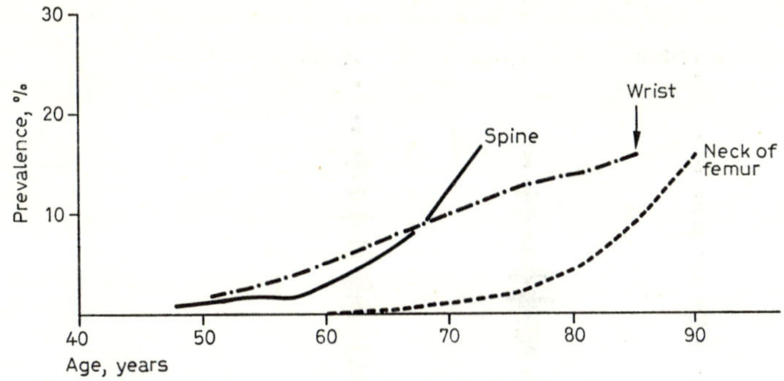

Fig. 2. Prevalence of wrist, spine and femoral neck fractures in women as a function of age. Note that the spine values must be an underestimate since known osteoporotics were excluded from the series.

about half of the former and 5% of the latter, the total skeletal loss between the ages of 50 and 80 in the average woman is about 15% [TROTTER *et al.*, 1960].

This loss of bone is associated with a rise in fracture rates, more particularly of the wrist, spine and femoral neck, which ultimately affect a substantial proportion of the female population. Figure 2 shows that some 15% of women who survive to the age of 75 will have had a wrist fracture, a similar proportion a femoral neck fracture [KNOWELDEN *et al.*, 1964] and a comparable proportion a vertebral crush fracture [SMITH and RIZEK, 1966]. If there was no coincidence between these three manifestations of osteoporosis, it would imply that something like 50% of women surviving to the age of 75 would have suffered an osteoporotic fracture. The true figures are somewhat less, however, since the fractures at these three sites do not all occur in different subjects. According to the data of ALFFRAM [1964] and our own data, about 20% of femoral neck fracture patients have suffered an earlier wrist fracture, and our own collection of spine X-rays in femoral neck fracture cases shows that about 40% of them have vertebral collapse. About 20% of our crush fracture cases have had a wrist fracture. If we adjust our total fracture prevalence to allow for this degree of coincidence – which can only be very approximate – it would seem that perhaps 30% of women reaching the age of 75 will have experienced one or more osteoporotic-type fracture.

To say that these are osteoporotic-type fractures is to imply that the reduced amount of bone has produced a critical decrease in bone strength

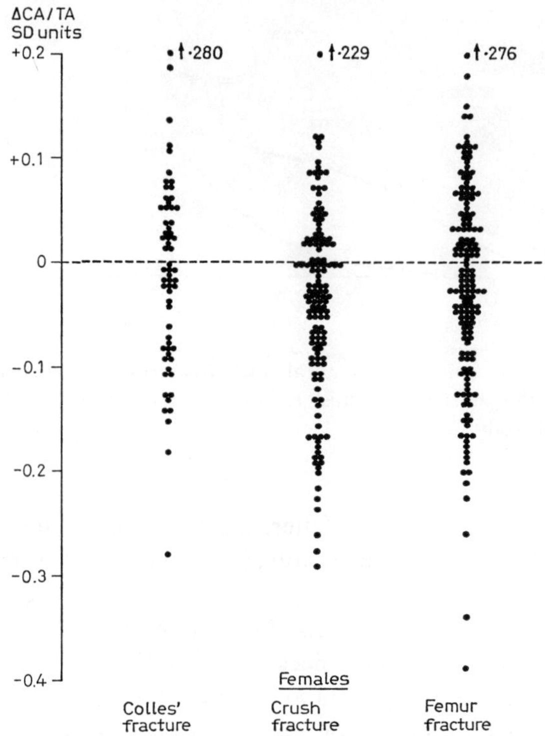

Fig. 3. Metacarpal area/total area ratio, corrected for age, in post-menopausal women *presenting* with wrist fractures, vertebral fractures and femoral neck fractures.

and significantly increased the fracture risk. It would be expected from this model that fracture cases as a group would have less bone than age-matched controls, and this is in fact correct. If the metacarpal cortex is taken as an indicator of bone status, it is clear that the wrist, spine and femoral neck fracture populations have significantly reduced CA/TA ratios for their age (fig. 3).

Pathogenesis

We have previously reported that corticosteroid-induced osteoporosis mainly affects post-menopausal women [GALLAGHER *et al.,* 1973b]. We have therefore compared normal, osteoporotic and steroid-treated post-menopausal women in respect of several measured variables.

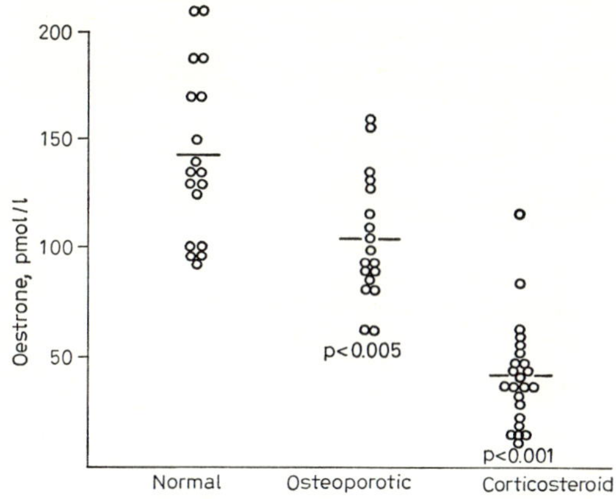

Fig. 4. Plasma oestrone levels in normal, osteoporotic and corticosteroid-treated post-menopausal women. The normal and osteoporotic cases have been matched for years since menopause.

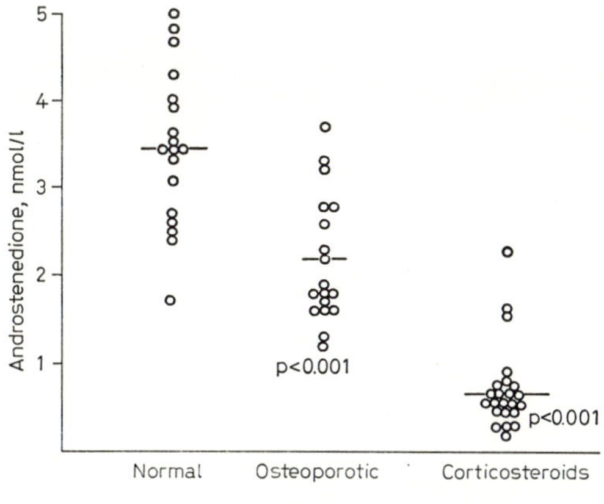

Fig. 5. Plasma androstenedione levels in the same patients as figure 4.

The plasma oestrone levels in the three groups of patients are shown in figure 4. Plasma oestrone is significantly lower in the osteoporotic than the normal group and significantly lower still in the steroid-treated cases. The corresponding androstenedione levels are shown in figure 5. Again, the plasma androstenedione is significantly reduced in the osteoporotics and significantly lower still in the steroid-treated cases. The corresponding plasma testosterone levels are shown in figure 6. The mean value is lower

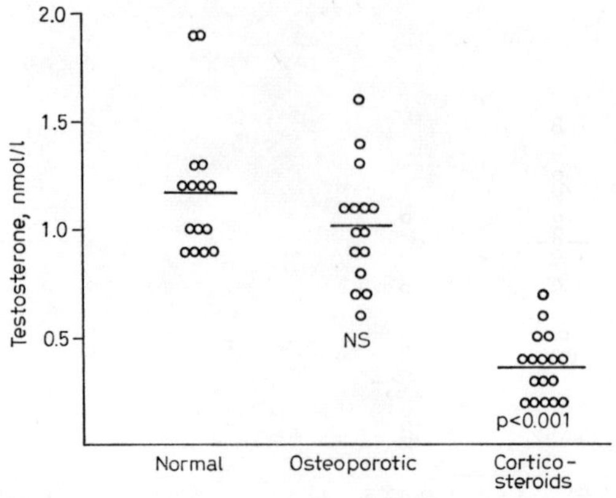

Fig. 6. Plasma testosterone levels in normal, osteoporotic and corticosteroid-treated post-menopausal women. The first 2 groups have been matched for years since menopause.

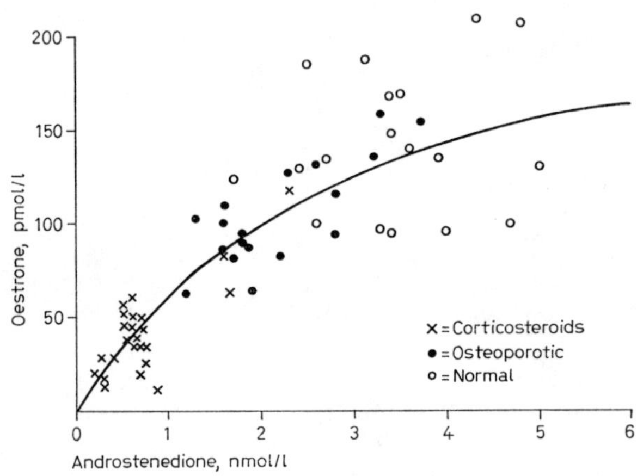

Fig. 7. The relation between plasma androstenedione and oestrone levels in the same patients as figures 4 and 5.

in osteoporotics than normals, though not significantly, and significantly lower still in the steroid-treated group.

The relation between the plasma androstenedione and oestrone levels in these three groups is shown in figure 7, together with the curvilinear regression of plasma oestrone on plasma androstenedione which we are describing elsewhere. The data suggest that the reduced plasma oestrone levels in the osteoporotic and corticosteroid-treated groups are the *result* of

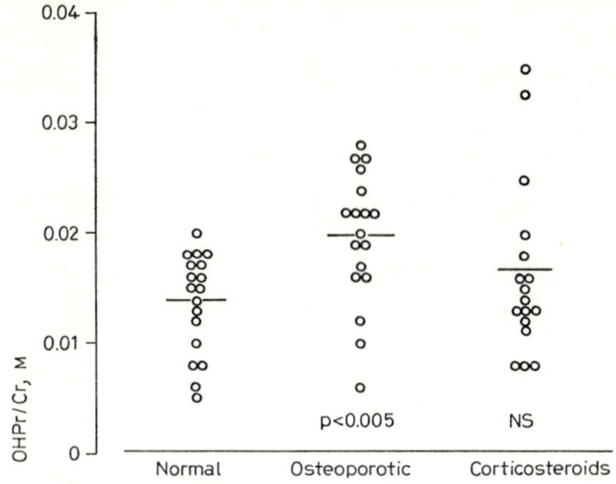

Fig. 8. Fasting urinary hydroxyproline and creatinine ratios in the same cases as figures 4 and 5.

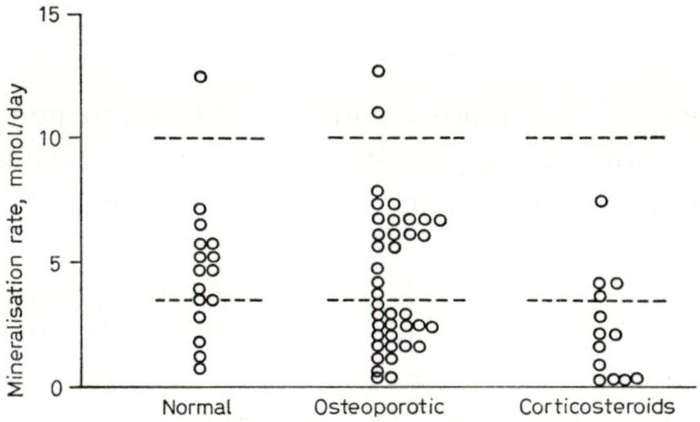

Fig. 9. Bone mineralisation rates in normal, osteoporotic and corticosteroid-treated post-menopausal women. The horizontal lines indicate the normal range in pre-menopausal women.

the low plasma androstenedione levels, as it is known that the main source of oestrone in post-menopausal women is by peripheral conversion of androstenedione [GRODIN *et al.*, 1973].

The fasting urinary hydroxyproline/creatinine ratios in the same three groups are shown in figure 8. Urinary hydroxyproline is significantly higher in the osteoporotics than the normals, indicating increased bone resorption in osteoporosis, but very variable in the corticosteroid-treated cases. Bone

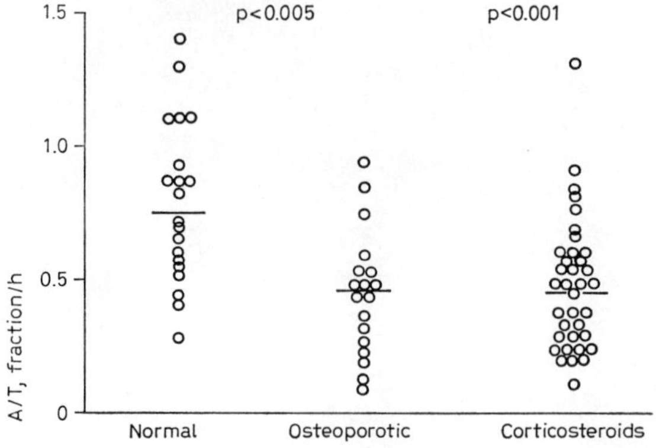

Fig. 10. Radiocalcium absorption in normal, osteoporotic and corticosteroid-treated post-menopausal women. The 'normal' cases in this figure are patients presenting with backache but found not to be osteoporotic who have been matched for years since menopause with the same osteoporotic cases as are illustrated in figures 3 and 4.

mineralisation rates in normal, osteoporotic and corticosteroid-treated post-menopausal women are shown in figure 9. Most of the normal cases fall within the range 3.5–10 mmol of calcium daily, which is our pre-menopausal range, but there are several cases below this range. About half the osteoporotic cases have mineralisation rates below 5 mmol/day and there is a strong suggestion that these data are bimodally distributed; thus there seem to be two osteoporotic populations as far as bone mineralisation rate is concerned. The majority of the steroid-treated cases fall below 5 mmol/day and resemble in this respect the group of osteoporotics with the low mineralisation rates.

Despite the wide variation in mineralisation rates in these cases, the calcium balance (i. e. the difference between the mineralisation and resorption rates) is very similar in each group. The normal group, the osteoporotic groups with normal and low mineralisation rates and the steroid-treated group are all in negative calcium balance, i. e. resorption exceeds mineralisation. An important factor in this negative calcium balance is malabsorption of calcium which is a feature of both the osteoporotic and steroid-treated cases as shown in figure 10.

The implication of these data is that low plasma hormone levels and malabsorption of calcium are probably both risk factors in the genesis of

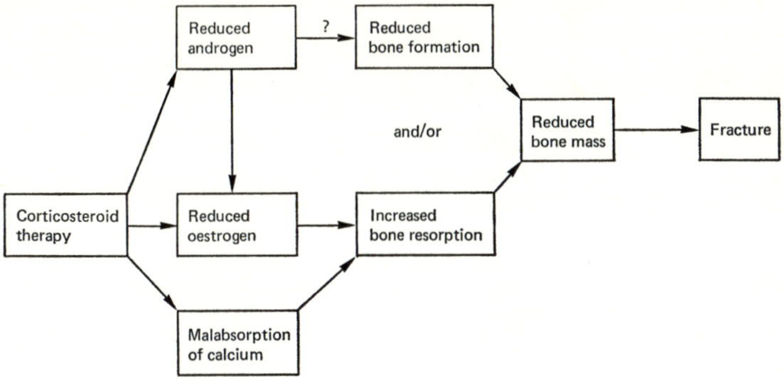

Fig. 11. Tentative scheme to illustrate the risk factors which lead to osteoporosis.

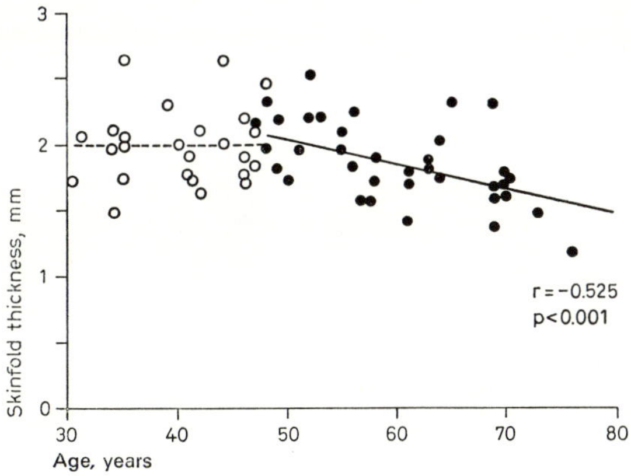

Fig. 12. Relation between skinfold thickness and age in pre-menopausal women over the age of 30 and post-menopausal women.

the bone loss which leads to osteoporosis. A tentative scheme showing how these risk factors lead to osteoporosis is shown in figure 11.

Skinfold Thickness

It is a well-known clinical observation that the skin is thin in patients with Cushing's syndrome and in those on corticosteroid therapy and it has also been claimed that the skin is 'transparent' in osteoporosis [McConkey *et al.*, 1963]. We have therefore measured skinfold thickness in normal, osteoporotic and corticosteroid-treated pre- and post-menopausal women. Figure 12 shows how skinfold thickness falls with age in post-menopausal

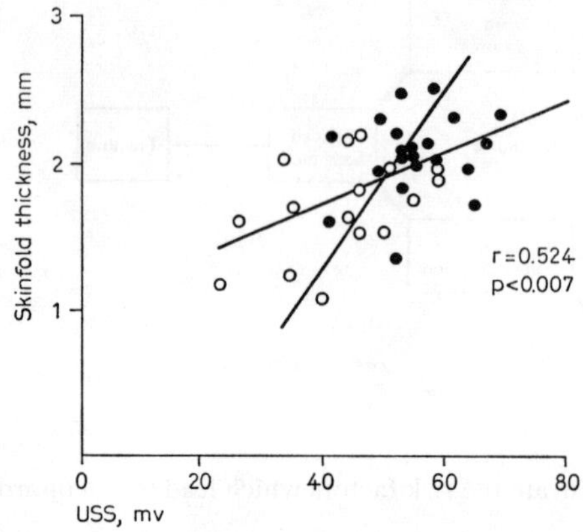

Fig. 13. The relation between skinfold thickness and urinary sediment smear maturation value (measured on the same day) in untreated (○) and oestrogen-treated (●) post-menopausal women.

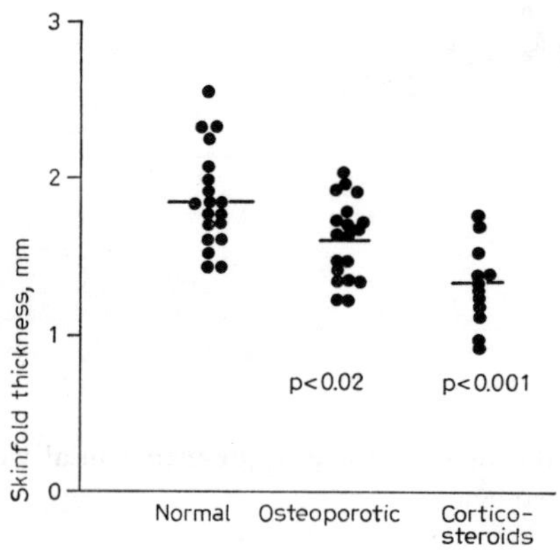

Fig. 14. Skinfold thickness in normal, osteoporotic and corticosteroid-treated post-menopausal women. The first 2 groups are matched for years since menopause.

women and figure 13 shows that skinfold thickness and maturation value appear to be related – at least when untreated and oestrogen-treated cases are combined. Moreover, as figure 14 shows, skinfold thickness is significantly reduced in osteoporotic patients and further reduced in those on corticosteroid therapy. In fact, skinfold thickness and bone mineralisation rate are significantly correlated as shown in figure 15. The patients with mineralisation rates below 5 mmol/day generally have a skinfold thickness

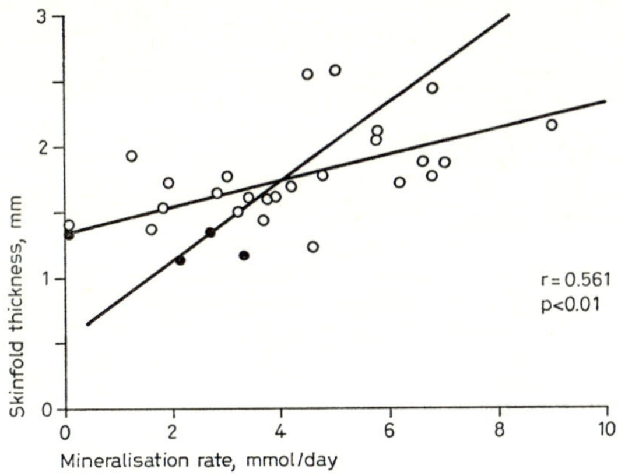

Fig. 15. The relation between skinfold thickness and bone mineralisation rate in normal, osteoporotic and steroid-treated (●) post-menopausal women.

below 1.7 mm and vice versa. In fact, there seems to be a general relationship between plasma androstenedione, skin thickness and bone mineralisation rate as shown in the table. Although the relationship between androstenedione and mineralisation rate is not significant, the data are very suggestive; the other two relationships are significant on the Chi-squared test.

Prevention and Treatment

We have previously reported that oestrogen therapy prevents post-menopausal bone loss and that calcium therapy retards it [HORSMAN *et al.*, 1977a, b]. Further data are presented in figure 16 which is based on the cases in our previous prospective trial and additional data now available to us from our Menopause Clinic. The treated and untreated cases have been divided into two matched groups according to the time that has elapsed since menopause, both because AITKEN *et al.* [1973] have reported that oestrogen therapy is ineffective if it starts more than 3 years after oophorectomy and because we have found accelerated bone loss in the first 3 years after this operation [HORSMAN *et al.*, 1977c]. These data show once again that oestrogen therapy prevents post-menopausal bone loss, and may even produce slight bone gain. The therapy appears to be equally effective whether it is given within 3 years of the menopause or later. Oestrogen therapy is also effective in preventing further bone loss in cases of established spinal osteoporosis *with normal calcium absorption* (fig. 17) but

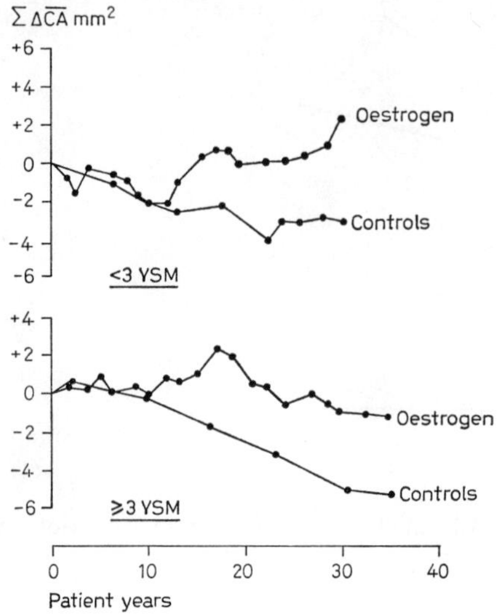

Fig. 16. Cusum plots of sequential changes in mean metacarpal cortical area in oestrogen-treated and untreated post-menopausal women.

cases with malabsorption require supplementary therapy with one of the vitamin D metabolites [MARSHALL and NORDIN, 1977].

Although oestrogen therapy clearly reduces bone resorption in normal and osteoporotic subjects, and may also be indicated in corticosteroid osteoporosis because of the low plasma oestrone levels reported above, there is no reason to suppose that it will increase the bone mineralisation rate which is a feature of so many of the osteoporotic cases and most of our steroid-treated cases. In fact oestrogen therapy tends, if anything, to reduce bone mineralisation rate [GALLAGHER and NORDIN, 1975] presumably because mineralisation and resorption rates are coupled. It is therefore unlikely that the low bone mineralisation rates we have described in half our osteoporotic cases can be attributed to oestrogen deficiency. It is possible, however, that androgens may be involved in the regulation of bone formation, and that there may be a case for androgen therapy in some of these patients, more particularly those on corticosteroids. Since androstenedione is the precursor of both oestrone and testosterone in postmenopausal women, elevation of the plasma androstenedione level presents an attractive therapeutic possibility. We have attempted this by administration of androstenedione, which, however, had a very short half-life

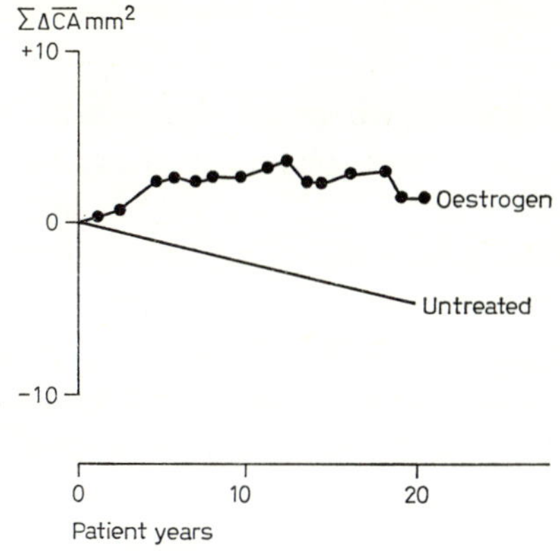

Fig. 17. Crush fracture cases: sex 3 + 4, all YSM. Cusum plot of sequential changes in mean metacarpal cortical area in oestrogen-treated and untreated cases of spinal osteoporosis.

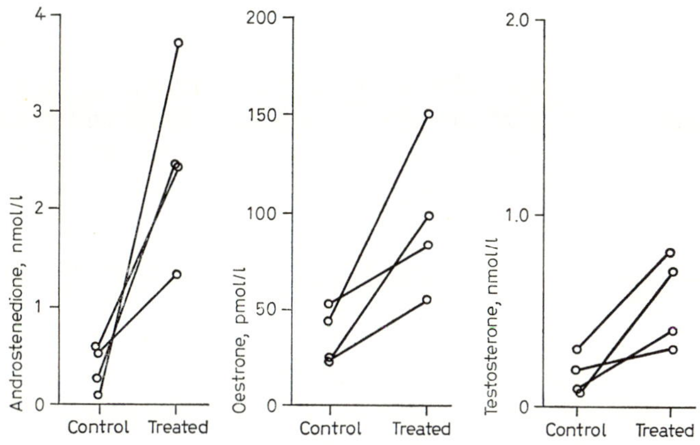

Fig. 18. Plasma androstenedione, oestrone and testosterone levels in 4 post-menopausal women on corticosteroid therapy before and during treatment with di-androne 10 mg b. d.

and are now looking one step further back in the metabolic pathway at dehydepiandrosterone, which is an important precursor of androstenedione itself. We have given this in the form of diandrone (supplied by Organon Laboratories) and determined the effect on plasma androgen and oestrogen levels in 4 cases as shown in figure 18. It is clear that this therapy raises the plasma levels of androstenedione, oestrone and testosterone.

Discussion

The relationhip between post-menopausal bone loss and fracture incidence is still not fully defined, but the broad picture is clear. Loss of bone increases the fracture risk and the fracture risk increases as the bone loss progresses. The wrist and vertebral fractures appear first and the femoral neck fracture later. Trabecular bone loss is probably the predominant factor in the first two fractures and is compounded by cortical bone loss in the femoral neck fracture – which is why it appears last. Vitamin D deficiency and secondary hyperparathyroidism-induced cortical bone resorption may be a significant additional factor in the femoral neck fracture since a high proportion of these patients have histological evidence of osteomalacia [AARON *et al.*, 1974]. Although there is some coincidence between these three fractures, i. e. wrist fracture cases have an increased risk of vertebral fracture and vertebral fracture cases an increased risk of femoral neck fracture, the majority of affected subjects sustain only one of these three fractures and the total proportion of the female population affected probably reaches something like 30% by the age of 70. Post-menopausal osteoporosis is therefore a major medical and social problem.

We have now shown that the plasma androstenedione and oestrone levels – and probably also the testosterone level – are reduced in spinal osteoporosis. The low oestrone and testosterone levels are almost certainly the result of the low plasma androstenedione levels, which must reflect either reduced adrenal secretion or failure of stromal androstenedione secretion by the post-menopausal ovary. The latter explanation is perhaps the less likely of the two because oophorectomy does not appear to increase the risk of osteoporosis to the degree that would be expected if the condition was due to ovarian insufficiency. It is true that about 15% of our osteoporotic cases have previously undergone bilateral oophorectomy, and this is probably a higher percentage of oophorectomy than in the population at large, but we have records of many oophorectomised women at all ages without fractures of any kind and it seems unlikely that ovarian stromal insufficiency *per se* is the explanation for our data in osteoporosis.

It is clear from our steroid-treated cases, with their extremely low oestrone and androstenedione levels, that the adrenal is normally the principal source of post-menopausal androstenedione, and therefore indirectly of oestrone, and that reduction in plasma androstenedione levels is therefore more likely to be adrenal than ovarian insufficiency. However, we have no evidence that our osteoporotic patients are suffering from ACTH or cortisol insufficiency, and if there is an adrenal androstenedione secretion failure, it is likely to be in the ratio of androstenedione to corticoids secreted by the adrenal. There is some evidence that prolactin may influence the corticoid/androgen ratio of adrenal secretion [BASSI *et al.*, 1977] and since prolactin also activates the one-hydroxylase in the kidney responsible for the synthesis of $1{,}25(OH)_2D_3$ [SPANOS *et al.*, 1976] the plasma level of which may be reduced in osteoporosis [GALLAGHER *et al.*, 1976] it would be an attractive concept if spinal osteoporosis were a prolactin deficiency. We are therefore looking at prolactin levels at the present time.

The fall in skin thickness after the menopause which we have described may reflect a fall in collagen content and arise mainly from a reduction in the dermis. However, the possibility that the epidermis shrinks after the menopause cannot be excluded [SHAHRAD and MARKS, 1976] and an increase in epidermal thickness on oestrogen has been reported by RAURAMO and PUNNONEN [1973]. We have no idea whether the reduced skin thickness in osteoporosis and in patients on corticosteroid therapy is epidermal or dermal. The structure of the vaginal epithelium is of course very similar to that of the skin, and it would be tempting to think that the dramatic effect of oestrogens on the vaginal epithelium has its counterpart in some effect on the epidermis as suggested by VAN KEEP [1973]. However, it is possible that skin thickness is more related to androgen than oestrogen status and that the apparent link between skin thickness and bone mineralisation rate, which we have reported, stems from a common androgen effect. There is a suggestion in our data that bone mineralisation rate may be related to the plasma androstenedione level and we are considering the possibility that androgens may regulate bone formation and oestrogens bone resorption; the combination of reduced androgen and oestrogen levels which occurs at the menopause and is accentuated in osteoporotic and steroid-treated cases may be an important factor in the development of bone loss. These concepts are reminiscent of the ideas of ALBRIGHT and REIFENSTEIN [1948], but he thought that oestrogens as well as androgens regulated bone formation and discounted the direct effect of these hormones on bone resorption.

Our data have considerable implications as far as the general use of oestrogens is concerned. It would not be wise to advocate universal hormone replacement therapy to protect a minority of the population from fractures, although our latest calculations suggest that this minority is in fact a very substantial one. However, our data do strongly suggest that the osteoporotic patients can be identified in a variety of ways – hormonal status, calcium absorption, urinary hydroxyproline and so on – and all our evidence suggests that these disturbances must precede the clinical onset of the disease by many years. We are therefore approaching a situation where it should be possible to identify the potential fracture cases by a variety of screening procedures and to recommend hormone therapy in this group of the population. All the oestrogens we have tested so far in ordinary replacement therapy doses are effective in reducing bone resorption. We are now trying to establish what is the lowest dose which has this effect, and the possible therapeutic role of androgen/oestrogen precursors.

References

AARON, J. E.; GALLAGHER, J. C.; ANDERSON, J.; STASIAK, L.; LONGTON, E. B., and NORDIN, B. E. C.: Frequency of osteomalacia and osteoporosis in fractures of the proximal femur. Lancet *i:* 229–233 (1974).

AITKEN, J. M.; HART, D. M., and LINDSAY, R.: Oestrogen replacement therapy for prevention of osteoporosis after oophorectomy. Br. med. J. *iii:* 515–519 (1973).

ALBRIGHT, F. and REIFENSTEIN, E. C.: The parathyroid glands and metabolic bone disease (Williams & Wilkins, Baltimore 1948).

ALFFRAM, P. A.: An epidemiological study of cervical and trochanteric fracture of the femur in an urban population. Acta orthop. scand *34:* suppl. 65, pp. 1–109 (1964).

BASSI, F.; GINSTI, G.; BORSI, L.; CATTANEO, S.; GIANNOTTI, P.; FORTI, G.; PAZZAGLI, M.; VIGIANI, C., and SERIO, M.: Plasma androgens in women with hyperprolactaemic amenorrhoea. Clin. Endocr. *6:* 5–10 (1977).

BURKINSHAW, L.; MARSHALL, D. H.; OXBY, C. B.; SPIERS, F. W.; NORDIN, B. E. C., and YOUNG, M. M.: Bone turnover model based on a continuously expanding exchangeable calcium pool. Nature *222:* 146–148 (1969).

CASTILLO, DEL E. B.; ARGONZ, J., and MAININI, C. G.: Cytological cycle of the urinary sediment and its parallelism with the vaginal cycle. J. clin. Endocr. *8:* 76–87 (1948).

GALLAGHER, J. C.; AARON, J.; HORSMAN, A.; MARSHALL, D. H.; WILKINSON, R., and NORDIN, B. E. C.: The crush fracture syndrome in postmenopausal women. Clin. Endocr. Metab. *2:* 293–316 (1973a).

GALLAGHER, J. J.; AARON, J.; HORSMAN, A.; WILKINSON, R., and NORDIN, B. E. C.: Corticosteroid osteoporosis. Clin. Endocr. Metab. *2:* 355–368 (1973b).

GALLAGHER, J. C. and NORDIN, B. E. C.: Oestrogens and calcium metabolism; in VAN KEEP and LAURITZEN Ageing and estrogens. Front. Hormone Res., vol. 2, pp. 98–117 (Karger, Basel 1973).

GALLAGHER, J. C. and NORDIN, B. E. C.: Effects of oestrogen and progestogen therapy on calcium metabolism in postmenopausal women; in VAN KEEP and LAU-

RITZEN Estrogens in the postmenopause. Front. Hormone Res., vol. 3, pp. 150–176 (Karger, Basel 1975).

GALLAGHER, C.; RIGGS, L.; EISMAN, J.; ARNAUD, S., and DELUCA, H.: Impaired intestinal calcium absorption in postmenopausal osteoporosis. Clin. Res. *24:* 360A (1976).

GRODIN, J. M.; SIITERI, P. K., and MACDONALD, P. C.: Source of estrogen production in postmenopausal women. J. clin. Endocr. Metab. *36:* 207–214 (1973).

HORSMAN, A.: Bone mass; in NORDIN Calcium phosphate and magnesium metabolism, pp. 357–404 (Churchill/Livingstone, Edinburgh 1976).

HORSMAN, A.; GALLAGHER, J. C.; SIMPSON, M., and NORDIN, B. E. C.: Prospective trial of oestrogen and calcium in postmenopausal women. Br. med. J. *ii:* 789–792 (1977a).

HORSMAN, A.; NORDIN, B. E. C.; GALLAGHER, J. C.; KIRBY, P. A.; MILNER, R. M., and SIMPSON, M.: Observation of sequential changes in bone mass in post-menopausal women: a controlled trial of oestrogen and calcium therapy. Calcif. Tiss. Res., suppl. 22, pp. 217–224 (1977b).

HORSMAN, A.; SIMPSON, M.; KIRBY, P. A., and NORDIN, B. E. C.: Non-linear bone loss in oophorectomized women. Br. Radiol. *50:* 504–507 (1977c).

KNOWELDEN, J.; BUHR, A. J., and DUNBAR, O.: Incidence of fractures in persons over 35 years of age. Br. J. prev. soc. Med. *18:* 130–141 (1964).

MARSHALL, D. H. and NORDIN, B. E. C.: The effect of 1αOHD$_3$ with and without oestrogens on calcium balance in postmenopausal women. Clin. Endocr., suppl. (in press, 1977).

MCCONKEY, B.; FRASER, G. M.; BLIGH, A. S., and WHITELEY, H.: Transparent skin and osteoporosis. Lancet *i:* 693–695 (1963).

NORDIN, B. E. C.; GALLAGHER, J. C.; AARON, J. E., and HORSMAN, A.: Post-menopausal osteopenia and osteoporosis; in VAN KEEP and LAURITZEN Estrogens in the post-menopause. Front. Hormone Res., vol. 3, pp. 131–149 (Karger, Basel 1975).

NORDIN, B. E. C.; HORSMAN, A., and AARON, J.: Diagnostic procedures; in NORDIN Calcium, phosphate and magnesium metabolism, pp. 469–524 (Churchill/Livingstone, Edinburgh 1976).

RAURAMO, L. and PUNNONEN, R.: The effects of castration and peroral estrogen therapy on a woman's skin; in VAN KEEP and LAURITZEN Ageing and estrogens. Front. Hormone Res., vol. 2, pp. 48–52 (Karger, Basel 1973).

SHAHRAD, P. and MARKS, R.: The effect of estrogens on the skin; in CAMPBELL The management of the menopause and post-menopausal years, pp. 243–251 (MTP, Lancaster 1976).

SMITH, R. W. and RIZEK, J.: Epidemiologic studies of osteoporosis in women of Puerto Rico and Southeastern Michigan with special reference to age, race, national origins and to other related or associated findings. Clin. Orthop. *45:* 31–48 (1966).

SPANOS, E.; PIKE, J. W.; HAUSSLER, M. R.; COLSTON, K. W.; EVANS, I. M. A.; GOLDNER, A. M.; MCCAIN, T. A., and MACINTYRE, I.: Circulating 1α,25-dihydroxy vitamin D in the chicken: Enhancement by injection of prolactin and during egg laying. Life Sci. *19:* 1751–1756 (1976).

TROTTER, M.; BROMAN, G. E., and PETERSON, R. R.: Densities of bones of white and Negro skeletons. J. Bone Jt Surg. *42:* 50–58 (1960).

VAN KEEP, P. A.: Discussion following RAURAMO and PUNNONEN: The effects of castration and peroral estrogen therapy on a woman's skin; in VAN KEEP and LAURITZEN Ageing and estrogens. Front. Hormone Res., vol. 2, p. 53 (Karger, Basel 1973).

Prof. B. E. C. NORDIN, Medical Research Council Mineral Metabolism Unit, The General Infirmary, Great George Street, *Leeds LS1 3EX* (United Kingdom)

Discussion

VAN KEEP: Is androstenedione effective in the *in vitro* model with a rat calvarium and PTH?

NORDIN: Yes it is. The molar concentrations are very high though. I do not know the physiological meaning of this.

LAURITZEN: Have estrogen receptors been found in bone?

THIJSSEN: We have not found them [PAASSEN, H. C. VAN; BORGART-CREUTZBURG, I.; POORTMAN, I.; THIJSSEN, J. H. H., and DUURSMA, S. A.: Calcif. Tiss. Res., in press, 1978]; neither have others. There has been a report on corticosteroid receptors in bone, however, and it may well be that some of the differences observed between the effects of corticosteroids and those of estrogens, are due to the fact that there are receptors for one and not for the other.

NORDIN: There does indeed appear not to be an estrogen receptor, but there may, of course, be many hormone actions on target tissues without a specific receptor being involved. There is a parathyroid hormone receptor in bone and, we know that estrogens do something – it may be on the Krebs cycle, it may be on ATP/ADP – which interferes with the action of the parathyroid hormone. I am sceptical about the role of the corticosteroid receptors. I am even beginning to doubt whether cortisone, certainly in the dose ranges used in clinical practice, has a direct effect on bone at all. Cortisone, in a bone tissue culture, blocks bone resorption. It seems not to be the direct effect of the steroids themselves – though I do not want to exclude this possibility – but the indirect effect through the suppression of the pituitary and the adrenals and, of course, the effect of corticosteroids on calcium absorption from the gut.

BAULIEU: I disagree with you. If there is an effect, there must, by definition alone, be a receptor system. I do not know of any instance where this is not so. It may be that there are purely technical reasons which prevent your being able to find it at the present time. There is evidence that there is an androgen receptor in bone. This receptor can, of course, also bind estrogens, whereas the opposite is not true, i. e., the estrogen receptor does not bind androgen. The affinity of this androgen receptor towards estrogens is not high, but the levels of circulating estrogens are higher when estrogens are administered than in physiological circumstances.

MYHRE: I should like to ask three questions: Is there a difference in body weight between osteoporotic and non-osteoporotic patients? If there is, one might suppose that this is because of the difference in androstenedione-estrone conversion. Have you observed a lower incidence in endometrial cancer in osteoporotic women? What is the role of fluoride in the treatment of osteoporosis?

NORDIN: The body weight of our osteoporotic patients is not significantly different from that of aged-matched controls. We have done some 60 conversion studies in osteoporotic and normal women and have not found a difference in the conversion rate. The difference is in the androstenedione value itself. The estrone level depends on the level of plasma androstenedione and on the conversion rate, but our belief is that osteoporosis is due to low androstenedione levels and not to any fault in the conversion.

As far as endometrial cancer is concerned, in the 10 years that we have been working in this field we have not seen a single case. To your third point: I have done no work at all on fluoride. I think that it is possible, however, that fluoride protects the bone against resorption by parathyroid hormone, in the same way that estrogens do.

VOIGT: When orchidectomized men suffering from prostatic carcinoma are treated with prednisolone there is a sharp fall in blood testosterone and estrone levels. There is convincing evidence that this results from the suppression of the adrenal cortex. When such patients are treated with estradiol or other estrogens there is an increase in the adrenocortical secretion, probably because of the stimulation of the ACTH production. It would be very interesting to know if this also happens when women are treated with estrogens.

NORDIN: I have indeed noticed when going through our data, that many patients on estrogen therapy have a raised androstenedione level in addition to a raised estrone level. Yours might indeed be the right explanation.

THIJSSEN: There is a paper from an Italian group reporting an increase of ACTH secretion in young women being treated with estradiol [GENAZZANI, A. R.: Hormone Res. *6:* 299–300 (1975)].

LAURITZEN: Your work on DHEA is interesting. I believe there is a case for giving DHEA to post-menopausal patients; the production rate, its level in the blood, and the excretion go down after 55 years of age. DHEA is an anti-catabolic hormone which is always secreted together with corticosteroids in stress situations, perhaps to counteract some of the unwanted effects of corticosteroids.

THIJSSEN: I do not think that there is a very direct relationship between cortisol and DHEA secretion. The daily variations do suggest this, but if you look into this over a long period of time – say 10–20 years – there is quite a discrepancy between the two. Because of this a separate factor, apart from ACTH, has been presumed, which controls DHEA secretion by the adrenal.

NORDIN: I take your point, but when ACTH is administered both the DHEA and the cortisol go up. Maybe prolactin is somehow involved?

BAULIEU: In the pharmacological department at Cambridge in the U. K., a hormonal factor exerting a positive control of DHEA adrenal production has been shown to exist in the human fetus. This factor is of pituitary origin. Has the conversion from DHEA to androstenedione been studied?

NORDIN: No. We have studied the conversion of androstenedione to estrone, but not of DHEA to androstenedione.

BAULIEU: Do your data suggest that the conversion from androstenedione to estrone is more efficient in osteoporotic women than in others?

NORDIN: No. We presume that somewhere in the conversion from androstenedione to estrone, there is a limiting step, a saturable stage in one of the enzymes or carriers. On that basis you get a line approaching asymptotically a maximum value of plasma estrone, when the rate-limiting enzyme-mediated step is saturated. It follows from this that the lower the substrate concentration the higher the percentage that is converted. This may explain why SIITERI and MACDONALD have found a higher conversion rate in post-menopausal than in pre-menopausal women: post-menopausal women have a lower plasma androstenedione, and therefore, a higher conversion rate. I have to admit that so far we have not been able to show convincing differences in the conversion rates between those on and off steroid therapy, but, of course, the error in determining the conversion rate is very substantial.

VOIGT: We have found the same picture in the conversion of testosterone to 5α-dihydrotestosterone (5α-DHT). The lower the testosterone values the higher the conversion rate to 5α-DHT. This may be explained by the mechanism outlined by you, but there may also be another explanation, i. e., the higher affinity to SHBG of 5α-DHT than of testosterone.

NORDIN: The similarity is striking, but your explanation for the conversion of testosterone to 5α-DHT cannot hold for that of androstenedione to estrone as androstenedione is not bound to SHBG.

KOPERA: Your recent experiments with castrated rats and with rats fed on a low calcium diet suggest that the greatest bone loss is seen when these two factors are combined. Were the data of this group significantly different from those of the oophorectomy-only group?

NORDIN: They certainly were. It was surprising to see how little effect oophorectomy alone had. In this connection, however, it is important to remember that the normal diet of the laboratory rat is extremely high in calcium – it is about half a gram per day, which is about the same as that of humans – which meant that we were in fact looking at oophorectomized rats on a high calcium diet and non-oophorectomized rats on a low calcium diet. The calcium deficiency was seen to be the more important factor, but, as you say, the greatest bone loss was seen when the two factors were combined. This was visible on the femoral cortical width, on the trabecular bone of the tail vertebrae, on the bone weight and on the ash weight of the bone.

KOPERA: Do you have any experience with the so-called anabolic steroids in addition to the estrogen?

NORDIN: No, I have not done work with anabolic steroids. We have done bone formation rate studies in male osteoporotics on testosterone. Testosterone is the only compound we have that puts up the bone formation rates. We are now studying the effect of ACTH on bone formation in women who were on corticosteroids before. It looks as though this also puts up the bone formation rate, perhaps because androstenedione and testosterone go up. Of course, in our immediate programme we are concentrating on DHEA. We are more interested in DHEA and androstenedione than in synthetic steroids.

VOIGT: Are you sure that it is testosterone and not 5α-DHT that is active in stimulating the bone formation rate?

NORDIN: I really cannot answer that. I do not know the fate of testosterone in the body.

HASPELS: Do you have any data on the effect of progestagens on bone?

NORDIN: We only have experience with norethisterone, a little of which is, of course, converted to estrogens. It has the same effects as estrogens on bone resorption. In fact I use 5 mg norethisterone for women in whom estrogen therapy is contra-indicated.

THIJSSEN: The conversion of norethisterone to estrogens is based on older reports, but in 1970 BREUER from Bonn suggested that the majority of this conversion is due to an artifact during the working-up of the urine. Only a minute part of norethisterone – much less than 0.1 % – is converted to estrogen [BREUER, H.: Lancet *ii:* 615–616 (1970)].

HASPELS: I usually give 5 mg of the progestagen lynestrenol per day in those rare cases where estrogens are contra-indicated or where estrogens cause unacceptable side effects. Its effects on hot flushes etc. are satisfactory, but you have to give local estrogens to correct an atrophy of the vaginal wall.

NORDIN: Norethisterone not only controls hot flushes, but the vaginal smear also reacts well. It looks as though you are giving an estrogen. Whatever the figures may show, I have a strong suspicion one is in fact giving a small amount of estrogen.

VAN KEEP: I think the Maturation Value goes up, but that you do not reach a complete maturation of the vaginal epithelium.

NORDIN: This is probably so, but you certainly get a significant rise in maturation value.

BAULIEU: The fact that progestagens like norethisterone are also active on the bone level confirms my point. It is likely that the androgen receptor probably present in bone is similar to all other androgen receptors. It binds estrogens and progestagens when these are administered in pharmacological dosages. Their effect may be due to an interference with the androgenic apparatus of the bone cells.

NORDIN: I do not think that the doses of estrogens we use should be called 'pharmacological doses'. I agree that they are probably larger than need be, and that the resulting plasma levels of estrone are tremendous, but the case for an effect on bone is not just based on that. It is based on the difference between pre- and post-menopausal women. You cannot call pre-menopausal women's estrogen levels 'pharmacological'.

BAULIEU: An unpublished report of KING in London establishes an interesting relation between the usual dose of Premarin used in post-menopausal women and the high estrogen status at the uterine level.

UTIAN: I would like to come back to the role of the post-menopausal ovary. About 10 years ago PROCOPÉ showed that, as differentiated by hormone studies and by histochemical staining of the ovary, there are two types of post-menopausal women: the 'producers' and the 'non-producers' [Acta endocr., Copenh., *60:* suppl. 135 (1968)]. More recently, VERMEULEN has shown that at least 50 % of the plasma testosterone and 30 % of the androstenedione of the post-menopausal woman is contributed by the ovary [J. clin. Endocr. Metab. *42:* 247 (1976)]. One would expect, therefore, an increase in the incidence of osteoporosis in post-menopausal women who were castrated.

Secondly, should we take plasma androstenedione measurements prior to surgery and, depending on the findings, decide which patients that should be castrated at the time of hysterectomy and which should not?

NORDIN: Our data on this have been published [MARSHALL, D. H.; CRILLY, R. G., and NORDIN, B. E. C.: Brit. med. J. *ii:* 1177 (1977)]. There is a very significant difference in plasma androstenedione and estrone levels between oophorectomised and non-oophorectomised normal women matched for years since the menopause; in osteoporotic non-oophorectomised women the levels are even lower than in the oophorectomised women without osteoporosis. It may look from this as though the estrogen deficiency syndrome of the post-menopausal woman who develops osteoporosis is basically an ovarian stromal failure. If this were so, oophorectomised women would have a very high risk of osteoporosis because they have got no ovarian stroma at all. ALBRIGHT reported that 15% of his osteoporotic women were oophorectomised, and we have exactly the same figure, but nearly 15% of our non-osteoporotic women are also oophorectomised, so we do not find a significant association between oophorectomy and osteoporosis. I think, therefore, that it is not an ovarian stromal problem. I think it is more likely to be an adrenal one. We are now doing dexamethasone suppression tests. We are giving dexamethasone for 48 h to switch off the adrenal and then measuring the androstenedione and estrone levels to see if the osteoporotic women are getting less androstenedione from their ovaries than the non-osteoporotic women. In the pre-menopause, about 40% of the androstenedione is ovarian in origin. There is a sharp drop when a woman goes through the menopause, and, in our experience, only 10–20% of the post-menopausal plasma androstenedione comes from the ovaries; even this is time limited. 3–4 years after the menopause there are still some women with relatively high androstenedione levels, for which I think the ovarian stroma is responsible. 10 years after the menopause all the androstenedione levels are down to about the same level.

UTIAN: I am not sure that one can be so dogmatic about this. Firstly, the dexamethasone suppression test can partially suppress the ovary. Secondly, one cannot discount VERMEULEN's work; he really did show a significant ovarian androgen production.

NORDIN: There is a significant contribution from the ovaries. There is no question about that. The fact remains, however, that the percentage of oophorectomised women is the same among osteoporotics as among non-osteoporotics.

MCKAY HART: I think I can add some evidence to support Professor NORDIN's point. Some years ago we studied bone density by photon absorption technique in some 120 patients in about half of whom the ovaries had been removed at hysterectomy whilst in the others they had been conserved. We found in those who had been operated upon at the age of 45 or less, there was a significant difference in the bone density between the oophorectomized women and those whose ovaries had been conserved. When the operation had taken place after the age of 45 there was no significant difference in bone density between the two groups, which suggests that after the age of about 45 the ovaries are not doing much to preserve bone density. The two groups of patients were pretty well age matched.

NORDIN: A final comment: Evolution has determined that the neuromuscular system is more important than the skeleton to survival. Therefore, the maintenance of the plasma calcium takes precedence over the maintenance of the integrity of the skeleton, and what we observe in osteoporosis is a situation in which the bone is resorbed to maintain the plasma calcium at the expense of the skeleton. In the premenopausal woman the estrogen levels are such that the parathyroid hormone does not attack the bone unless it is absolutely essential for the maintenance of plasma calcium, as when the calcium intake is very low. In the post-menopausal woman the relative sensitivities of gut, bone, and kidney to the action of the parathyroid hormone [and/or 1.25 $(OH)_2D_3$] is so altered that when the parathyroid hormone acts in order to maintain the plasma calcium, it works directly on the bone, instead of upon the gut and the kidney, as it ought to do. The bone is all too easily resorbed. This at least is the hypothesis upon which we are working at present and with which all the data collected so far are compatible.

Front. Hormone Res., vol. 5, pp. 76–88 (Karger, Basel 1978)

The Morphology of Endometrial Proliferative Reactions

G. P. Vooijs

Institute of Pathology, Catholic University of Nijmegen, Nijmegen

Before venturing into the problem of evaluating the significance of morphologic changes in the endometrium under the influence of estrogens, one should ensure that the endometrial tissue fragments under study are representative for the entire endometrium. We shall therefore primarily discuss the diagnostic accuracy of several sampling methods.

The most important methods presently in use are: (1) dilatation and curettage; (2) endometrial biopsy (microcurettage); (3) suction biopsy (vabra aspirator); (4) endometrial lavage (Gravlee jet wash); (5) aspiration cytology (endometrial and endocervical aspiration).

Dilatation and Curettage

Meticulously carried out dilatation and conventional curettage under anesthesia gives an excellent and, in almost all instances, representative picture of morphologic changes in the endometrium. The overall accuracy of conventional curettage in experienced hands is 95% or better. The major source of unsatisfactory specimens is an atrophic endometrium. For a definitive diagnosis of precancerous or cancerous endometrial growth, especially when still limited in extent, it is our preferred method.

However, for screening for endometrial cancer, for follow-up studies and for women who should not be exposed to general anesthesia, one should carefully judge the advantages of less traumatic methods such as endometrial biopsy, vacuum curettage, endometrial lavage and aspiration cytology, either used individually or in combination when several methods might increase the detection rate.

Table I. Diagnostic accuracy of endometrial biopsy

Author	Year	Patients	Accuracy in cancer detection, %
Palmer et al.	1950	301 (95)	90
Jordan et al.	1956	128 (25)	92
McGuire	1962	103 (38)	85
Slaughter et al.	1962	406 (68)	76

The number of patients with endometrial cancer are indicated in parentheses. From COHEN *et al.* [1974].

Endometrial Biopsy

The endometrial biopsy was introduced into gynecologic practice as an outpatient procedure by KELLY in 1925 and has since been performed with a large variety of ingeniously designed instruments ranging from brushes, sponges, wires and bristles to small sharp curettes.

Although relatively little is known of the comparative diagnostic accuracy of endometrial biopsy and conventional curettage, some studies indicate that in cases of neoplasia the two procedures seem to be equally dependable (table I). The detection rate of endometrial cancer in representative studies varies from 76 to 92 %.

However, in view of the limited amount of tissue which can be obtained by the biopsy technique this procedure seems less reliable for the detection of localized lesions.

Suction Biopsy

Diagnostic vacuum curettage of the uterus was introduced by JENSEN and JENSEN in 1968. In comparative studies, several investigators [BJERRE *et al.*, 1971; SAUNDERS and ROWLAND, 1972; DENIS *et al.*, 1973; COHEN *et al.*, 1974] (table II) have tried to determine the diagnostic accuracy of the suction curettage method by comparing histopathologic diagnoses in patients who underwent a conventional curettage after a suction curettage.

In a successful aspiration sample one obtains a quantity of tissue comparable to that obtained in a conventional curettage. SAUNDERS and ROWLAND [1972] found that over 20 % of the specimens were unsatisfactory for histological diagnosis. The majority of failures, however, were due to faulty technique. DENIS *et al.* [1973] found an unsatisfactory rate of only 5 % and COHEN *et al.* [1974] an unsatisfactory rate of only 3 %. DENIS

Table II. Diagnostic accuracy of suction curettage

Author	Year	Patients	Accuracy, %
Bjerre et al.	1971	111 (2)	94
Saunders et al.	1972	73 (1)	80
Denis et al.	1973	134 (1)	97.8
Cohen et al.	1974	98 (1)	95

The number of patients with endometrial cancer are indicated in parentheses.

et al. [1973] found a positive correlation of histologic diagnoses between specimens in as much as 97.2% of the satisfactory specimens.

Based on results obtained in postmenopausal women, this method is considered inadequate when endometrial atrophy is present. Vacuum curettage causes stromal edema and cytoplasmic vacuolization which may lead to the erroneous conclusion of secretory activity. In cases of early hyperplasia the loosening of the cellular stroma may leave, as an only criterion, minor cystic changes of the glands which can then lead to a false-negative diagnosis. The main sources of inaccuracy seem to be the inability to enter the cervix and the presence of endometrial polyps. Each of the authors employing the vacuum technique has missed polyps.

With advanced and generalized lesions this method may be used with success, but in cases of localized lesions, as in iatrogenic estrogen-induced hyperplasia, this procedure may prove to be less adequate. However, suction curettage is still the preferred method for outpatient curettage screening.

Gravlee Jet Irrigation

Soon after the introduction of the irrigator by GRAVLEE [DOWLING *et al.*, 1969; GRAVLEE, 1969] a number of studies were carried out to assess the diagnostic accuracy of this uterine lavage technique [UNDERWOOD *et al.*, 1973; BIBBO *et al.*, 1974; RODRIQUEZ *et al.*, 1974; SMITH *et al.*, 1974; TWIGGS *et al.*, 1976]. Based on these studies we can conclude that the irrigation method fails to detect a high percentage of hyperplastic and localized or deeply buried premalignant and malignant lesions and that it is not sufficiently reliable to exclude with certainty the presence of a malignant process even when a negative sample is obtained. Most studies have a high percentage of unsatisfactory specimens (2.5%–27%) which may be caused by such factors as active bleeding, inflammations and necrosis of the endometrium which leads to physiologic and irregular shedding.

If we accept that endometrial carcinoma is the final result of a gradually increasing dedifferentiation of endometrial glandular cells, it seems unlikely that the Gravlee jet wash will accurately identify at an early stage those high-risk patients in whom a malignant process is destined to develop [Twiggs *et al.*, 1976].

Aspiration Cytology

Samples obtained directly from the endometrial cavity consistently contain a large concentration of cells in an optimal state of preservation. However, this time-consuming procedure is technically difficult and requires sterile conditions. Aspiration from the endocervical canal is the best alternative to endometrial aspiration [Reagan and Ng, 1973]. Such a readily obtained sample contains an adequate number of desquamated superficial epithelial as well as endometrial glandular cells.

The evaluation of endometrial cells in endocervical aspiration samples requires an experienced cytopathologist. Since the diagnosis is based on exfoliated cells accumulated in the endocervical canal the more dedifferentiated processes are more readily detected. This is illustrated in the detection rates for precursors of endometrial cancer [Ng *et al.*, 1973; Ng, 1974] and grades I and II endometrial cancer in comparison to grades III and IV endometrial cancer (table III).

Table III. Cellular detection of endometrial carcinoma and endometrial carcinoma precursors

Histology	Cellular detection, %
Cystic hyperplasia	67.9
Adenomatous hyperplasia	70.3
Atypical hyperplasia	75.3
Adenocarcinoma *in situ*	80.0
Adenocarcinoma grade I	64
Adenocarcinoma grade II	78
Adenocarcinoma grade III	81
Adenocarcinoma grade IV	83
Adenoacanthoma	76.0
Adenosquamous carcinoma	92.5

From Ng [1974].

In experienced hands an overall accuracy in the detection of endometrial cancer and its precursors of 75% can be obtained. This detection rate warrants the routine use of this procedure in a population at risk. Furthermore, the study of exfoliated or imprinted single cells as an additional part of a histopathological investigation often adds detailed information on the cellular and nuclear level and may ultimately lead to a better understanding of the significance of minute cellular changes in the process of dedifferentiation of endometrial epithelial cells.

The human endometrium is an end organ which is highly sensitive to stimulation with gonadal steroids. The physiologic action of estrogens on the endometrium is a proliferation of epithelial and stromal cells of the superficial endometrial layers in cyclic women, and is known as the proliferative phase of the menstrual cycle. The tissue is composed of single round or oval glands surrounded by a densely packed stroma. As a sign of active growth in both structures, numerous mitoses indicating cell division can be easily found. Administration of estrogens during the proliferative phase leads to prolonged proliferation. If unopposed by progestogens this estrogen-mediated proliferative activity may lead to an excessive overgrowth of endometrial elements eventually resulting in the development of endometrial cancer.

The carcinogenic potency of endogenous estrogens is well established through the increased incidence of endometrial cancer in women with ovarian cystic syndrome (Stein-Leventhal) and estrogen-producing ovarian tumors.

Proliferating endometrium

↓

Glandular cystic hyperplasia

↓

Adenomatous hyperplasia

↓

Atypical adenomatous hyperplasia

|

Adenocarcinoma *in situ*

↓

Adenocarcinoma

Fig. 1. Proliferative changes of the endometrium

Table IV. Endometrial carcinoma

Morphology of non-involved endom.	Premenopausal (53 patients)		Postmenopausal (147 patients)	
	n	%	n	%
Cystic gl. hyp.	23	43.4	51	34.0
Atyp. endom. hyp.	26	49.0	40	27.2
Benign polyps	14	26.4	41	27.9

From GRAY *et al.* [1974].

There is now considerable evidence that exogenous estrogens as well are linked with an increased risk of endometrial cancer [ZIEL and FINKLE, 1975; MACK *et al.*, 1976; GORDON *et al.*, 1977; GRAY *et al.*, 1977].

By now it is generally accepted that the majority of endometrial carcinomas develop through successive stages of endometrial hyperplasia. Continuous high doses of estrogens may exhaust cellular enzymes and make the target tissue less susceptible for these hormones, finally resulting in atrophy.

If the levels of estrogen remain elevated for a longer period endometrial hyperplasia may develop into an adenocarcinoma in susceptible patients. This is characterized by an overgrowth of glands and ever-increasing cellular dedifferentiation. This hypothesis is strongly supported by the finding of all successive stages of endometrial hyperplasia coincidentally with adenocarcinoma in the noninvolved parts of the endometrium [GRAY *et al.*, 1974] (table IV).

Sustained action of endogenous estrogens or the prolonged administration of small doses of estrogen may induce the endometrium to undergo a glandular cystic hyperplasia. The level of the estrogen concentration is less important than the duration of its unopposed (by progestogens) effect. Glandular cystic hyperplasia is most common between the ages of 41–50 when the ovarian function is in transition and estrogen production fluctuates.

Hyperplasia may have a unicentric or multicentric focus of origin. Diffuse involvement of the endometrium may occur. However, the malignant potential of glandular cystic hyperplasia is low: only 14.6% of untreated women with a glandular cystic hyperplasia ultimately developed an adenocarcinoma [VELLIOS, 1972]. The most common type is the homologous endometrial hyperplasia in which glands and stroma proliferate concurrently. The glands do not increase in number but proliferate intensely and undergo cystic dilatation.

The glands are usually lined by tall columnar and sometimes ciliated cells with evidence of mitoses. The nuclei are usually basal and occasionally intermediate in location. There is abundant stroma separating the dilated glands. The stroma cells may be increased in size with frequent mitoses. Occasionally a secretory type of endometrial hyperplasia may be found. Synthesis of DNA in nuclei of proliferating glandular cells is often increased. DNA values of glandular cystic hyperplasia and most of the adenomatous hyperplasias are unequivocally diploid [WAGNER and TERNER, 1967]. The stimulating effect of synthetic estrogens on glandular elements differs from the effect on stromal cells [DALLENBACH-HELLWEG, 1975].

In 2.5 % of hyperplasia specimens islands of squamous epithelium are found. These are either derived from pluripotential Müllerian cells or from metaplasia of glandular columnar cells. DALLENBACH-HELLWEG [1975] regards these cells as an irregularity in the reaction of the endometrium on prolonged elevated levels of estrogens. This may occur in all stages of hyperplasia eventually resulting in the development of an adenoacanthoma [DALLENBACH-HELLWEG, 1975].

The duration of the hyperestrogenism determines what happens to the glandular cystic hyperplasia. If the levels of estrogen fall within the onset of menopause regressive changes may occur and a cystic atrophy ensues. This change might be regarded as a petrified state of hyperplasia [DALLENBACH-HELLWEG, 1975]. If the hyperestrogenism persists after menopause, glandular cystic hyperplasia may progress to adenomatous hyperplasia. Proliferating glandular epithelial cells cause a relative overgrowth of stroma cells. Hyperestrogenism becomes manifest by the appearance of foam cells occurring in about 50 % of cases. The number of these foam cells is considered an indicator of the prognostic course of the hyperplasia. The more numerous the foam cells the sooner an adenomatous endometrium progresses to a carcinoma [DALLENBACH-HELLWEG and RUDOLPH, 1974].

Microscopic protrusions of the proliferating glands into the surrounding stroma form satellite nests of closely packed glands. The glandular cells may be slightly larger than regular proliferating cells and may show some stratification. The nuclei usually have a basal, sometimes an intermediate position and are slightly enlarged. The nuclear chromatin is finely granular. Micronucleoli may be found but are infrequent.

Atypical hyperplasia is a more advanced stage and the glandular abnormalities are more closely related to those of adenocarcinoma. The lesion is characterized by large glands which lie closely together but are still separated by definite stroma strands. The glandular epithelial cells are frequently enlarged and pseudostratified. The glandular lining epithelium may show papillary infoldings.

The cytoplasm is amphophilic or basophilic. The nuclei are enlarged and often differ in size. They may have a basal or intermediate position and show stratification and sometimes a slightly altered polarity. The nuclear chromatin is coarsely granular and unevenly distributed. Mitoses are common. Increased polymorphism of glandular cells usually is not evenly distributed within the glandular epithelium. A total of 75 % of women followed with an atypical adenomatous hyperplasia eventually developed adenocarcinoma [VELLIOS, 1972]. The hyperplasias developing after many years of exogenous estrogen therapy are characterized by special morphological features. They develop multicentrally and even in the adenomatous stage may be circumscribed or polypoid. The cells comprising the adenomatous regions show a great variation in size and shape. Flattened and nodular areas of squamous (metaplastic) epithelium are common [DALLENBACH-HELLWEG, 1975]. The differential diagnosis between atypical hyperplasia and adenocarcinoma often poses great difficulties and may cause great disagreement between investigators. However, an experienced pathologist who applies strict morphologic definitions may achieve considerable agreement on most cases, as was demonstrated by MCKAY *et al.* [1956] when reviewing the endometrial biopsies from the original ZIEL and FINKLE study [GORDON *et al.*, 1977]. In a small number of cases there is still no means of distinguishing between atypical hyperplasia and adenocarcinoma with any degree of confidence [NOVAK and WOODRUFF, 1974]. These cases may constitute a gradual transition between the two lesions. This distinction is probably of little clinical importance since the biologic behavior of these two lesions is almost similar and the subsequent therapy and prognosis of early endometrial carcinoma and atypical adenomatous hyperplasia are similar [GORDON *et al.*, 1977]. Differences in statistics on the malignant potential of these lesions can often be based on differences in morphological criteria on which a diagnosis of atypical hyperplasia or adenocarcinoma is based.

Adenocarcinoma *in situ* is generally regarded as the most advanced step prior to the development of cancer. It arises often multifocally in areas of adenomatous hyperplasia [BUEHL *et al.*, 1964], and is considered to be a preinvasive carcinoma. Microscopically it is characterized by localized proliferation of glands.

There is no invasion of the endometrial stroma or the myometrium. The glands tend not to touch but remain separated by strands of stroma cells. The glands are usually larger than their normal counterparts. Papillary infoldings and reduplications of the glandular epithelium are conspicuous. Stratification of the cells is common. The epithelium lining the glands is of tall columnar type. Cellular polarity may be altered. The cytoplasm of the cells of the adenocarcinoma *in situ* is strikingly pale and clear because they contain only small amounts of RNA [McKAY *et al.,* 1956]. The cells have abundant amphophilic or basophilic cytoplasm. The enlarged round to oval nuclei have a basal, intermediate or superficial position and have a coarsely granular, somewhat increased, unevenly distributed chromatin. Micronucleoli are frequent but macronucleoli are uncommon. Mitoses are infrequent. Cytophotometric studies reveal an aneuploidy like that found in invasive cancer [WAGNER and TERNER, 1967]. Adenocarcinoma *in situ* may coexist with adenomatous or atypical hyperplasia and the lesion may be preceded by either of these changes. It appears certain that the biologic outcome of this change if unopposed by progestogens is an adenocarcinoma [WENTZ, 1973].

Ultimately adenocarcinoma develops. It may be localized or diffuse. Several different types of endometrial cancer are recognized. The more common types of neoplasms of the endometrium are: (1) adenocarcinoma; (2) adenoacanthoma (3) clear cell carcinoma, and (4) adenosquamous carcinoma.

Adenocarcinoma

Essential for an adenocarcinoma is the loss of the normal endometrial pattern. Glands are increased in number and closely packed with little or no intervening stroma. The glands are usually somewhat smaller than their normal counterparts and variable in size and shape. The lining epithelium may be simple columnar or pseudostratified with a variable loss of polarity. Mitoses are frequently found. Well-differentiated adenocarcinomas tend to show a glandular pattern. With progressive dedifferentiation the glandular pattern becomes less prominent and a papillary pattern or a solid growth pattern without acinar formation is more often observed.

Cells of an adenocarcinoma are larger than normal endometrial epithelial cells and may be columnar, cuboidal or irregular in form. Nuclei are increased in size and vary somewhat in shape [REAGAN and NG, 1973]. Nuclear chromatin is coarsely granular and unevenly distributed. Nucleoli are present and tend to be macronucleoli in the more dedifferentiated forms. GRAY *et al.* [1977] could not observe a difference in the stage of the endometrial carcinomas between estrogen and nonestrogen users. Although not statistically significant there was a larger number of well-differentiated cancers among estrogen users.

Adenoacanthoma

An endometrial adenocarcinoma having a malignant glandular component and a benign squamous component. The squamous nodules in an adenoacanthoma may represent remnants of metaplastic epithelium that arose in pre-existing benign hyperplasia. They appear commonly after years of estrogen therapy. The glandular component of an adenoacanthoma is not significantly different from that of other adenocarcinomas. Adenoacanthomas comprise about 25% of endometrial adenocarcinomas [REAGAN, 1974a, b].

Clear Cell Carcinoma

Clear cell carcinoma is an uncommon variety of endometrial adenocarcinoma. The tumor is composed of large columnar or polyhedral cells which show an abundant clear cytoplasm. Nuclei tend to be eccentrically located (hobnail cells). Macronucleoli are usually prominent. These neoplasms are morphologically similar to the clear cell carcinomas found in cervix and vagina. Although a relation between intrauterine exposure to synthetic estrogens and vaginal or cervical clear cell carcinomas has been established no such relation has been found for endometrial clear cell carcinomas.

Adenosquamous Carcinoma

An endometrial carcinoma having a malignant squamous and a malignant glandular component. This carcinoma seems to show a relative increase in incidence lately and tends to occur at a later age (average 64 years). A correlation between the increased use of synthetic estrogens and increased incidence of adenosquamous carcinoma has as yet not been established.

This concludes the discussion of the proliferative reactions of the endometrium during exogenous estrogen medication. Apart from the apparently multicentric origin, as yet no specific characteristics on cellular or nuclear level of these induced proliferative changes have been identified. Since our knowledge of these iatrogenic changes is still very limited much work has to be done before we can exclude such characteristic features.

References

BIBBO, M.; RICE, A. M.; WIED, G. I., and ZUSPAN, F. P.: Comparative specificity and sensitivity of routine cytologic examinations and the Gravlee jet wash technique for diagnosis of endometrial changes. Obstet. Gynec. *43:* 253–256 (1974).

BJERRE, B. A.; GARDMARTE, S., and SJOBERG, N.: Aspiration curettage – a new diagnostic method. J. Reprod. Med. *7:* 221–223 (1971).

BUEHL, I. A.; VELLIOS, F.; CARTER, J. E., and HUBER, C. P.: Carcinoma *in situ* of the endometrium. Am. J. clin. Path. *42:* 594 (1964).

COHEN, C. J.; GUSBERG, S. B., and KOFFLER, D.: Histologic screening for endometrial cancer. Gynecol. Oncol. *2:* 279–286 (1974).

DALLENBACH-HELLWEG, G.: Histopathology of the endometrium; 2nd ed. (Springer, New York 1975).

DALLENBACH-HELLWEG, G. and RUDOLPH, E.: Foam cells and estrogenic activity of the human endometrium. Arch. Gynaek. *217:* 335 (1974).

DENIS, R.; BARNETT, J. M., and FORBES, S. E.: Diagnostic suction curettage. Obstet. Gynec. *42:* 301–303 (1973).

DOWLING, E. A.; GRAVLEE, L. C., and HUTCHINS, K. E.: A new technique for the detection of adenocarcinomas of the endometrium. Acta cytol. *13:* 496–501 (1969).

GORDON, J.; REAGAN, J. W.; FINKLE, W. D., and ZIEL, H. K.: Estrogen and endometrial carcinoma. Pathological support of original risk estimate. New Engl. J. Med. *297:* 570–571 (1977).

GRAVLEE, L. C.: Jet irrigation method for the diagnosis of endometrial carcinoma, its principle and accuracy. Obstet. Gynec. *34:* 168–172 (1969).

GRAY, L. A.; CHRISTOPHERSON, W. M., and HOOVER, R. N.: Estrogens and endometrial carcinoma. Obstet. Gynec. *49:* 358–392 (1977).

GRAY, L. A.; ROBERTSON, R. W., and CHRISTOPHERSON, W. M.: Atypical endometrial changes associated with carcinoma. Gynecol. Oncol. *2:* 93–100 (1974).

JENSEN, J. A. and JENSEN, J. G.: Vacuum curettage for diagnostic purposes. Ugeskr. Læg. *130:* 2124 (1968).

JORDAN, M. J.; BADER, G., and NEMAZIE, A. S.: Comparative accuracy of preoperative cytologic and histologic diagnoses in endometrial lesions. Obstet. Gynec. *7:* 646 (1956).

KELLY, H. A.: Curettage without anesthesia on the office table. Am. J. Obstet. Gynec. *9:* 78 (1925).

MACK, T. M.; PIKE, M. C., and HENDERSON, B. E.: Estrogens and endometrial cancer in a retirement community. New Engl. J. Med. *294:* 1262–1267 (1976).

MCGUIRE, T. H.: Efficacy of endometrial biopsy in the diagnosis of endometrial carcinoma. Obstet. Gynec. *19:* 105 (1962).

MCKAY, D. G.; HERTIG, A. T.; BARDAWIL, W., and VELARDO, J. T.: Histochemical observations on the endometrium. Abnormal endometrium. Obstet. Gynec. *8:* 140 (1956).

NG, A. B. P.: The cellular detection of endometrial carcinoma and its precursors. Gynecol. Oncol. *2:* 162–179 (1974).

NG, A. B. P.; REAGAN, J. W., and CECHNER, R. L.: The precursors of endometrial cancer, a study of their cellular manifestations. Acta cytol. *17:* 437–448 (1973).

NOVAK, E. R. and WOODRUFF, J. D.: Novak's gynecologic and obstetric pathology; 7th ed. (Saunders, Philadelphia 1974).

PALMER, J. P.; KNEER, W. J., and ECCLESTON, H. H.: Endometrial biopsy, comparison of aspiration curettage with conventional dilatation and curettage. Am. J. Obstet. Gynec. *60:* 671 (1950).

REAGAN, J. W.: The changing nature of endometrial cancer. Gynecol. Oncol. *2:* 144–151 (1974a).

REAGAN, J. W.: Endometrial carcinogenesis. Compendium on diagnostic cytology, vol. 3 (Tutorials of Chicago, Chicago 1974b).

REAGAN, J. W. and NG, A. B. P.: The cells of uterine adenocarcinoma; 2nd ed. (Karger, Basel 1973).

RODRIGUEZ, M. A.; RUBIN, A., and KOSS, L. G.: Evaluation of endometrial jet wash technique (Gravlee) in 303 patients in a community hospital. Obstet. Gynec. *43:* 392–399 (1974).

SLAUGHTER, C. R. and SCHEWE, E. J.: Evaluation of biopsy of the endometrium by the Novak suction curette. Am. J. Obstet. Gynec. *83:* 1302 (1962).

SAUNDERS, P. and ROWLAND, R.: Vacuum curettage of the uterus without anaesthesia. J. Obstet. Gynaec. Br. Commonw. *79:* 168–174 (1972).

SMITH, J. P.; EHRLICH, C. F., and LUKEMAN, J. M.: Detection of endometrial carcinoma using a jet washer. Obstet. Gynec. *43:* 522–526 (1974).

TWIGGS, L. B.; DISAIA, P. J.; MORROW, C. P.; TOWNSEND, D. E., and SCHWINN, C. E.: Efficacy in diagnosis of endometrial neoplasia. J. Am. med. Ass. *235:* 2748–2750 (1976).

UNDERWOOD, E. B.; KELLET, M. D., and MCKEE, E. E.: The Gravlee jet washer. Can it replace the diagnostic curettage? Am. J. Obstet. Gynec. *117:* 201–209 (1973).

VELLIOS, F.: Endometrial hyperplasias, precursors of endometrial carcinoma; in SOMMERS Pathol. Annu., pp. 201–229 (Appleton Century Crofts, New York 1972).

WAGNER, D. and TERNER, J. Y.: Desoxyribonucleic acid content of precursors of endometrial carcinoma. Cancer *20:* 1067 (1967).

WENTZ, W. B.: Progestin therapy in endometrial hyperplasia. New directions for research in endometrial cancer. Proc. Am. Cancer Soc. (1973).

ZIEL, H. K. and FINKLE, W. D.: Increased risk of endometrial carcinoma among users of conjugated estrogens. New Engl. J. Med. *293:* 1167–1170 (1975).

Prof. G. P. VOOIJS, Instituut voor Pathologische Anatomie, Katholieke Universiteit Nijmegen, Geert Grooteplein Zuid 24, *Nijmegen* (The Netherlands)

Discussion

UTIAN: In the United States, REAGAN has noted an increase in adenosquamous carcinoma, and suggested that the observed increase in the incidence of uterine cancer is to be explained by this rather than by an increase in the incidence of plain adenocarcinoma. He has also suggested that, in Cleveland at least, the increase in adenosquamous carcinoma runs parallel with the increase in estrogen use, and that these two might be related.

Vooijs: I have been working under Reagan. As far as I know, the subject is still under study, but the figures do suggest a relation. Others, however, have observed this increase in adenosquamous carcinoma, but have not found a relation with estrogen use.

Comninos: Could you comment on the reversibility of the changes you have described? Is there a spontaneous reversibility? How much time does it take to progress from one stage to the next: from atypical adenomatous hyperplasia to carcinoma *in situ* or invasive carcinoma?

Vooijs: As far as reversibility is concerned, it is a fact that many cases of hyperplasia do not progress at all. They may reverse at the time of menopause. With atypical adenomatous hyperplasia the situation is different: if untreated, 80% progresses in time into an adenocarcinoma. We know next to nothing about the speed of progress of such lesions.

Studd: Have you ever come across patients who had atypical hyperplasia that reversed under progestagen therapy? We had four worrying pathological Vabra biopsies, all of which were diagnosed by at least one pathologist as adenocarcinoma. Subsequently, with norethisterone, the pathology improved to a less sinister hyperplasia. 3 months later there was no sign of any pathology at all. Two of the patients received norethisterone continuously for 2 months, and other two received the progestagen in courses of 10 days with intervals. All four had either a complete amenorrhoea or normal cyclical bleedings before diagnoses were made, thus there was no clue in the menstrual bleeding pattern to the presence of endometrial pathology.

Vooijs: I have seen one case which reversed under progestagen therapy, but when the therapy was discontinued the adenomatous hyperplasia reappeared. The patient had, however, a cystic ovary. Eventually carcinoma of the endometrium developed.

The debate around the morphological features of the various types of endometrial pathology is only beginning, and at the moment resembles the one we had some 20 years ago around cervical pathology. I am sure that many cases of atypical hyperplasia are erroneously diagnosed as adenocarcinoma, and that these cases would reverse under progestagen therapy. Even an adenocarcinoma may reverse under such therapy, at least according to morphological criteria.

Van der Does: Do you see different results when using different methods of obtaining material, i. e., suction curettage such as the Vabra method, and the classical D and C?

Vooijs: I have no comparative data and I do not know of any, but it is my impression that with the Vabra one obtains material only from the surface of the endometrium. Some lesions, those originating in the basal layers of the endometrium, may not be found in material obtained by this method; they can only be found in material obtained with more force.

Front. Hormone Res., vol. 5, pp. 89–100 (Karger, Basel 1978)

The Epidemiology of Endometrial Cancer

A. Klopper and V. Farr

Department of Obstetrics and Gynaecology, University of Aberdeen, Aberdeen

The Incidence of Carcinoma Corpus Uteri

Although practitioners of epidemiology may not care for it, most dictionary definitions of the science involve the concept of prevalence in a community – which starts us on a controversial note. Is the incidence of uterine cancer changing? Does it differ from one race to another, or in different communities? About its changing rate there is some hard evidence worth looking at; about its distribution there are plenty of *ex cathedra* statements – that it is less common in Jewesses [3], or uncommon in Japan [23, 26].

The incidence of carcinoma of the endometrium appears to have begun to increase in the 1960s. This embracing statement requires modification and explanation. The earlier statistics are vitiated by the fact that many countries reported cancer of the cervix and cancer of the endometrium together as 'uterine cancer'. Despite their propinquity these two cancers involve quite different tissues, have a different aetiology and their incidence has probably been changing in opposite directions. Not surprisingly figures on 'cancer of the uterus' conceal more about endometrial cancer than they reveal. Although there had earlier been hints that endometrial cancer was on the increase, the definitive evidence was not forthcoming until 1976 when Weiss *et al.* [24] published an analysis of endometrial carcinoma from 8 areas in the USA. They found an increase 'of a magnitude that has rarely been paralleled in the history of cancer reporting'. Although there is evidence of a similar rise in other countries [17] all the authors are at one in attributing this to the increasing use of oestrogen replacement therapy in menopausal women [4, 16]. The role of oestrogen therapy in endometrial carcinoma will be considered later. Taking the assumption as correct for the moment, it is to be expected that the increase in endometrial cancer would first be noted in the USA as the oestrogen treatment of menopausal women is viewed with a good deal more enthusiasm there than elsewhere. It is also true that the proportion of postmenopausal women who have undergone hysterectomy in the United States is higher than in other countries, and this would conceal the real extent of the increase. The evidence for a dramatic increase in endometrial carcinoma is sound. The connection with oestrogen therapy may be less certain but is good enough to justify a critical reappraisal of current practice.

The Risk Factors

In 1947, CORSCADEN and GUSBERG [3] analysed the records of the Sloane Hospital for Women and contrasted the features of women with carcinoma of the corpus with those who had a cervical carcinoma. In the intervening 30 years we have added so little of substance to their analysis that it is worth trying to epitomise their findings in cartoons. Figure 1 shows the typical victim of cervical carcinoma – thin, poor and worn out by childbearing. Figure 2 shows the woman who gets a cancer of the corpus uteri – affluent, older, fat and nulliparous. We can pad this picture out a bit and add such factors as diabetes, hypertension, menstrual irregularity and delayed menopause. Two kinds of factors bear upon the risk of developing an endometrial carcinoma. One set, such as failure of ovulation, menstrual irregularity and a delayed menopause involves the oestrogen status of the woman. The other, constitutional and metabolic factors such as obesity and diabetes, appears to be independent, but often a case can be argued for their operation through the effects of oestrogen on the endometrium.

Constitutional Factors

Obesity

The list of authors who have noted that women with carcinoma of the endometrium tend to be heavier than other women of their age, is almost as long as the list of everybody who has written about the epidemiology of carcinoma corpus uteri. Such a widespread consensus lends weight to the

1

2

Fig. 1. The woman who gets a carcinoma of the cervix.
Fig. 2. The woman who gets a carcinoma of the endometrium.

observation, and indeed all that present-day epidemiology has to offer is to fill in the details; how much heavier, how strong is the association and above all how does fat operate to cause an endometrial cancer? Simply demonstrating that an unduly large proportion of women with carcinoma of the endometrium are heavier than is to be expected in normal women of their age, hardly establishes the connection between obesity and carcinoma corpus uteri. The more sophisticated analyses have involved control groups of various kinds. In this respect one of the most convincing studies was that of MUCK and TROTNOW [15]. They used two control groups, one of women presenting with postmenopausal bleeding but who were found to have an atrophic endometrium and the second control group a number of women admitted for pelvic floor repair. The average weight and the number of overweight women was significantly greater in the cancer group.

An ingenious explanation of why endometrial cancer is commoner in fat women has lately come to the fore [19]. In the postmenopausal woman, oestrogens are produced, not in the ovary, but by the peripheral conversion in body fat of androstenedione to oestrone. Oestrone is thus the predominant oestrogen in postmenopausal women just as oestradiol is in premenopausal. Oestrone acts somewhat differently from oestradiol on the target cells in the endometrium and is more apt to set in train the abnormal processes of cell division which result in cancer. The more fat, the more oestrone is produced. The more oestrone there is, the greater the liability to cancer. This thesis has been argued so persuasively that it is widely accepted. A few apparent contradictions give pause for thought. One of the most penetrating recent studies, albeit not specifically concerned with oestrone, found that oestrogen administration acts independently of obesity in causing endometrial carcinoma [22]. Siiteri and his colleagues [20] themselves found that the conversion of androstenedione into oestrone by fat women tended to increase, not decrease, when they lost weight.

Diabetes

The association of diabetes, or at least impaired carbohydrate tolerance, with carcinoma of the endometrium is not nearly as strong as that of obesity. Nevertheless, it is a recurring theme in many studies and is probably real. On the other hand, obesity and diabetes go together and it may be that the association of diabetes and endometrial carcinoma is a spurious one, resulting from the common factor of obesity. This possibility is discounted in the work of Muck and Trotnow [15] who found that obese women with endometrial cancer more often had diabetes than did obese controls. Wynder *et al.* [26] also concluded that high blood sugar was more common in women with endometrial carcinoma than could be accounted for by obesity.

Hypertension

Hypertension is coupled with diabetes in almost every analysis which mentions it [1, 14, 16, 18, 22, 26]. Only Dunn and Bradbury [5] found that neither hypertension nor diabetes occurred more frequently in their series of endometrial carcinoma cases than in the controls. Hypertension is common both in diabetics and in obese women. There is no convincing evidence that it has any direct connection on its own with endometrial carcinoma.

Affluence

One of the early papers on the epidemiology of endometrial carcinoma noted that many of these women were private patients in comfortable circumstances [8]. This is only to be expected in postmenopausal nulliparous women and the point was only raised for the sake of the contrast with the thin, younger woman worn out with childbearing who develops a carcinoma of the cervix. Once made, the point was perpetuated from paper to paper without thought or pertinent data. There is no reason to suppose that anything in the diet or the way of life of rich women makes them more susceptible to endometrial cancer.

Age

There is no doubt about carcinoma corpus uteri being a disease of postmenopausal women. A study done in Tübingen showed that only 2% of all cases of endometrial carcinoma occurred in women less than 40 years old [2], and other series showed an even smaller incidence of younger women. Of course it is difficult to find any kind of control group which has been age matched. In the few instances where this has been done, the age of the carcinoma patients has been significantly higher [10, 18]. The age of endometrial carcinoma patients tends to be brought down by virtue of the fact that some of these cancers arise in younger women with Stein-Leventhal syndrome. This may be deceptive. It has been argued that the tumour in young hirsute women with polycystic ovaries is a different entity from that which occurs in older women [25]. *Ex cathedra* statements that carcinoma corpus uteri is occurring in younger women have been made [2] but have not been backed up by any evidence.

Genetic Factors

Evidence of a familial factor in endometrial cancer has often been sought and occasionally found [15, 33]. If the connection was a strong one it would have been noted more often.

The Role of Oestrogens in the Genesis of Endometrial Cancer

Endogenous Oestrogens

The incidence of endometrial carcinoma is increased in a number of conditions which are presumed to be hyperoestrogenic states. The most often quoted association of endometrial cancer with excess oestrogen production is the high incidence of carcinoma corpus uteri in patients with an antecedent ovarian carcinoma. For the most part these are merely repetitions of a widespread belief but three authors have offered some original data [1, 7, 23]. It is of course very difficult to define any control group in this situation and these writers have not attempted to do so. Nevertheless, the figures are convincing and there is no reason to doubt that women with ovarian cancer have a high incidence of uterine cancer. It is less certain that the operative cause is high oestrogen production. For the most part the ovarian tumours concerned have been thecomas, not granulosa cell tumours. Although thecomas are often oestrogen-secreting tumours they also frequently produce androgens and it is just possible that the carcinogenic agent is an androgen, not an oestrogen. The same argument might also apply to the raised incidence of endometrial cancer in women with polycystic ovarian disease. JACKSON and DOCHERTY [10] found 16 cases of endometrial carcinoma in 43 women with Stein-Leventhal syndrome. They noted that the women with cancer were older and one wonders whether some of the younger women might not have gone on to develop an endometrial carcinoma. A somewhat different conclusion was drawn by WOOD and BORONOW [25]. On the basis of only two cases of polycystic ovarian disease with corpus carcinoma they reasoned that the tumour in women with polycystic ovarian disease was a different entity from the endometrial carcinoma of postmenopausal women. In the polycystic ovary cases the tumour occurred in young women and was associated with a low FSH and high LH secretion.

A further finding which is often cited in support of the role of endogenous oestrogen in endometrial carcinoma concerns adenomatous hyperplasia of the endometrium. There is some evidence of a causative association between high oestrogen levels and such hyperplasia [7]. The causative role of oestrogens with regard to endometrial hyperplasia has been somewhat more convincingly demonstrated in the case of exogenous oestrogen administration [6, 7]. Hyperplasia may precede a carcinoma and often coexists with it. Many pathologists conceive of adenomatous endometrial hyperplasia as a precancerous condition and indeed hyperplasia may shade into frank malignancy by such gradual progression that it is impossible to draw the line between them.

Other accepted epidemiological characteristics of endometrial cancer can be construed as expressions of hyperoestrogenicity. It might be supposed that the high incidence of nulliparity in endometrial carcinoma is due to anovulation. Anovulatory states are presumed to be situations of high oestrogen production, unopposed by progesterone. A similar assumption may be taken to explain the late onset of the menopause. Although infertility and the menopause are both undoubtedly associated with a failure of ovarian progesterone production, there is no reason to suppose that they are commonly marked by high oestrogen production. Indeed the more closely one comes to look at the role of endogenous oestrogen in endometrial carcinoma, the more does it appear that a case can be made for oestrogen unopposed by progesterone and possibly for prolonged oestrogen action but not for high oestrogen.

Taken together the factors of nulliparity, menstrual irregularity, adenomatous hyperplasia and delayed menopause spell aberrant ovarian steroidogenesis, not necessarily hyperoestrogenicity. The normal ovary produces oestradiol and progesterone, hirsute women with polycystic ovaries produce androgens, and fat old women produce oestrone and no progesterone.

Exogenous Oestrogens

The part played by oestrogen treatment of menopausal women as a cause of endometrial cancer has leapt into prominence in the last 2 years [2, 13, 27]. Although women using oral contraceptives take more oestrogen for a longer time, attention has been focussed, with one exception [21] on menopausal therapy. It is assumed, without convincing evidence, that young women on oral contraceptives are protected from endometrial carcinoma by the opposed action of the progestogens which they are also ingesting. None of the advocates of the protective effect of progestogens offer any definition of the manner in which progestogens 'oppose' oestrogens. There is no evidence that progestogens oppose oestrogens in the sense of competing for intracellular receptors. Oestrogens and progestogens set in train related, but separate, trains of intracellular events. By and large, progestogen-induced effects, particularly on the human endometrium, are dependent on the prior action of oestrogens and progestogens have no capacity for preventing effects which can be induced by oestrogens. An alternative possibility is that progestogens do not oppose oestrogens but that oestrogen-progestogen therapy involves periodic withdrawal bleeding which strips away the stimulated endometrium which might otherwise go on to become cancerous. This is quite a different kettle of fish.

Other authors in this volume have been set the task of examining the role of oestrogen therapy in the causation of endometrial carcinoma and it is not intended to trespass on their ground. One observation does properly fall within the province of the epidemiologist. This concerns the chemical nature of the oestrogen which causes the cancer. Oestrogens are very diverse in their structure and it makes no biological sense to lump them all together. As far as endogenous oestrogens are concerned, all the evidence is against one compound, oestrone [20]. With regard to exogenous oestrogens the evidence applies to oestrone sulphate and conjugated equine oestrogens [28], to synthetic oestrogens such as stilboestrol or ethynyl oestradiol or nonsteroidal oestrogens [7, 21, 28, 29]. There is no evidence inculpating the natural steroids, oestradiol-17β and oestriol. The latter, in particular, needs a closer look in view of its capacity for preventing the endometrial binding of other oestrogens [12].

References

1 ANDREWS, W. C.: Estrogens and endometrial carcinoma. Obstetl gynec. Surv. *16:* 747–767 (1965).

2 BACHMANN, F. F.: Das Adenokarzinom des Endometrion bei Frauen unter 40 Jahren. Zentbl. Gynäk. *95:* 1729–1742 (1973).

3 CORSCADEN, J. A. and GUSBERG, S. B.: The background of cancer of the uterus. Am. J. Obstet. Gynec. *55:* 419–431 (1947).

4 DOLL, R.; KINLEN, L. J.; SKEGG, D. L., and SMITH, P. G.: Hormone replacement therapy and endometrial carcinoma. Lancet *ii:* 745 (1977).

5 DUNN, L. T. and BRADBURY, J. T.: Endocrine factors in endometrial carcinoma. Am. J. Obstet. Gynec. *97:* 465–471 (1967).

6 GEIST, S. H. and SALMON, U. J.: Are estrogens carcinogenic in the human female? Am. J. Obstet. Gynec. *41:* 29–36 (1941).

7 GUSBERG, S. B.: Precursors of corpus carcinoma, estrogens and adenomatous hyperplasia. Am. J. Obstet. Gynec. *54:* 905–927 (1947).

8 GUSBERG, S. B.: Hormone dependence of endometrial cancer. Am. J. Obstet. Gynec. *30:* 287–293 (1967).

9 HOOVER, R.; EVERSON, R.; FRAUMEN, J., and MYERS, M.: Cancer of the uterine corpus after hormonal treatment for breast cancer. Lancet *i:* 885–887 (1976).

10 JACKSON, R. L. and DOCHERTY, M. B.: The Stein-Leventhal syndrome: analysis of 43 cases with special reference to the association with endometrial carcinoma. Am. J. Obstet. Gynec. *75:* 161–173 (1957).

11 LAU, H. U.; PETSCHELT, E.; POEHLS, H.; POLLEX, G.; UNGER, H., und ZEGENHAGEN, V.: Epidemiologische Aspekte beim Korpuscarcinom. Zentbl. Gynäk. *97:* 1025–1036 (1975).

12 LUCAS, W. E.: Causal relationships between endocrine-metabolic variables in patients with endometrial carcinoma. Obstet. gynec. Surv. *29:* 507–531 (1974).

13 MACK, T. M.; PIKE, M. C.; HENDERSON, B. E.; PFEFFER, R. I.; GERKINS, V. K.; ARTHUR, M., and BROWN, B. E.: Estrogens and endometrial cancer in a retirement community. New Engl. J. Med. *294:* 1262–1267 (1976).

14 MACMAHON, B.: Risk factors for endometrial cancer. Gynecol. Oncol. *2:* 122–129 (1974).

15 MUCK, B. R. und TROTNOW, S.: Endometriumcarcinom und pathologische Glucosetoleranz. Arch. Gynaek. *216:* 301–310 (1973).

16 QUINT, B. C.: Changing patterns in endometrial carcinoma. Am. J. Obstet. Gynec. *122:* 498–501 (1975).

17 ROSOL, M.; STRNAD, L.; HAVEI, V.; ANDRYS, J., und VACHA, K.: Untersuchungen zum Inzidenzanstieg der Endometrium- und Ovarialkarzinome in der CSSR während der Jahre 1960 bis 1973. Zentbl. Gynäk. *98:* 175–182 (1976).
18 SCHINDLER, A. E. und GLASER, F.: Untersuchungen zur Korpuskarzinom-Genese. Geburtsh. Frauenheilk. *34:* 186–194 (1974).
19 SIITERI, P. K. and MACDONALD, P. C.: Role of extraglandular estrogen in human endocrinology; in GREEP Handbook of physiology, vol. 11; Female reproductive system, pp. 615–629 (American Physiological Society, Washington 1973).
20 SIITERI, P. K.; WILLIAMS, J. E., and TAKAKI, N. K.: Steroid abnormalities in endometrial and breast carcinoma: a unifying hypothesis. J. Steroid Biochem. *7:* 897–903 (1976).
21 SILVERBERG, S. G. and MAKOWSKI, E. L.: Endometrial carcinoma in young women taking oral contraceptives. Obstet. Gynec. *46:* 505–506 (1975).
22 SMITH, D. C.; PRENTICE, R.; THOMPSON, D. J., and HERRMANN, W. L.: Association of exogenous estrogen with endometrial carcinoma. New Engl. J. Med. *293:* 1164–1167 (1975).
23 WAY, S.: The etiology of carcinoma of the body of the uterus. J. Obstet. Gynaec. Br. Commonw. *61:* 46–58 (1954).
24 WEISS, N.; SZEKELY, D. R., and AUSTIN, D. F.: Increasing incidence of endometrial cancer in USA. New Engl. J. Med. *294:* 1259–1262 (1976).
25 WOOD, C. P. and BORONOW, R. C.: Endometrial carcinoma and the polycystic ovary syndrome. Am. J. Obstet. Gynec. *124:* 140–142 (1976).
26 WYNDER, E. L.; ESCHER, G. C., and MANTEL, N.: An epidemiological investigation of cancer of the endometrium. Cancer *19:* 489–520 (1966).
27 ZIEL, H. K. and FINKLE, W. D.: Association of oestrone with the development of endometrial carcinoma. Am. J. Obstet. Gynec. *124:* 735–740 (1976).
28 ZIEL, H. K. and FINKLE, W. D.: Increased risk of endometrial carcinoma among users of conjugated estrogens. New Engl. J. Med. *293:* 1167–1170 (1975).

Prof. A. KLOPPER, Department of Obstetrics and Gynaecology, University of Aberdeen Medical School, Foresterhill, *Aberdeen AB9 2ZD* (United Kingdom)

Discussion

DE WAARD: I should like to make several comments. First of all, Dr. KLOPPER spoke only very briefly about the data relating to the incidence of endometrial carcinoma in various populations. I should have thought that there were now enough data from a sufficiently large number of countries for us to say that the incidence runs parallel to that of cancer of the breast, the colon, and the ovary.

Secondly, I should like to mention some work done in Johannesburg where a study was made of the development of endometrial cancer in the Bantu race. In the early fifties, although cervical cancer was quite common, endometrial cancer was extremely rare. It was seen, however, that as the traditionally lean rural-dwelling Bantu women came to live in towns they became fat, and, at the same time, despite the fact that they were definitely not taking estrogens, their smears became estrogenic. On the basis of this it was predicted that the incidence of endometrial carcinoma would increase, and this has indeed happened. Incidentally, it was also found that when the obesity was accompanied by hypertension or by diabetes the incidence of estrogenic smears increased considerably [WAARD, F. DE and OETTLE, A. G.: Cancer *18:* 450–459 (1965)].

Finally, we have found in a retrospective study that post-menopausal women with an estrogenic smear had a later menopause than other women. We presume that in these women there is a non-ovarian estrogen production on top of the normal ovarian one, and that it is this which delays the cessation of menstruation.

KLOPPER: I did not mean to imply that there were no data on racial and cultural differences; I only meant that these data have been rather recently obtained and, unlike diabetes, hypertension, and the other factors I mentioned in my review, have not yet found their way into the body of accepted criteria and knowledge on the subject. I am sure, however, that in the years to come, the study of these comparative data will be very rewarding.

I am a little hesitant as to what deductions one may draw from the Bantu study. This population undergoes tremendous changes as it moves from the rural areas to the big cities. So many things change and change so radically, that it is difficult to make deductions about changes in the incidence of carcinoma of the endometrium. Moreover, in the early fifties hardly any D and Cs were done on Bantu women, so the changes of an endometrial carcinoma being diagnosed were very remote.

DE WAARD: Because of the high incidence of cervical carcinoma many hysterectomies were done, and these uteri were studied.

KLOPPER: That is still a bias. I only want to say, however, that I hesitate to accept data from a population which is exposed to diagnostic measures at an accelerating rate. This may give false figures if you try to project backwards in time to conclude that the incidence has changed.

LAURITZEN: I understood that in the black population to which you are referring there is a high incidence of myomata, a condition which is statistically positively related to endometrial cancer. However, the observed incidence of endometrial cancer is low. This shows that correlations which are valid in one race or in one society are not necessarily valid in another.

KLOPPER: Professor DE WAARD's last point about estrogenic smears being more common in women with a late menopause is an intriguing one. Some 20 years ago, WAY made a similar observation [J. Obstet. Gynaec. Br. Commonw. *61:* 46–58 (1954)]. Professor DE WAARD's hypothesis for its explanation is a very valid one.

NORDIN: Are estrone levels in plasma of endometrial cancer patients different from those of age-matched controls? If we explain the influence of obesity by the increased conversion of androstenedione to estrone there should be a difference or our explanation is wrong.

THIJSSEN: JUDD found that there was no difference in estrone levels in plasma between endometrial cancer patients and age-matched, weight-matched controls [J. clin. Endocr. Metab. *43:* 272–278 (1976)].

KLOPPER: This does not necessarily mean that our explanation is wrong. The difference may be a very subtle one and then repeated essays on large numbers of women are going to be necessary in order to prove a difference. Also, measuring urinary estrone excretion or plasma estrone levels are crude, inexact ways of determining the estrone levels operative at the level of the endometrium. To a considerable extent estrone is being converted into estradiol and its sulphate. The fact that an obvious association is not easily found between estrone levels and endometrial carcinoma does not mean that a real connection between the two does not exist. It is a connection beset by many factors and the estrone differences may be so subtle as to be undetectable by our present means.

NORDIN: It may also be that there is no connection, and that there is some other cause.

BAULIEU: Professor KLOPPER mentioned the 'estrogen-opposing' effect of progesterone. In fact the situation is quite complex. For instance, progesterone itself does not interfere with estrogen receptors as it is not bound to them, but it does change the estrogen-receptor concentration. The mechanism of this is not yet completely clear. In the castrated adult rat, progesterone increases the estrogen receptors in the endometrium but not in the myometrium. Administered in the luteal phase of the cycle of a woman, progesterone decreases the number of estrogen receptors. I have my doubts about differences between various estrogens. All estrogens are able to enter the target cells, do, almost by definition, bind to the estrogen receptors, and trigger the mechanism leading to mitosis. This growth-promoting effect potentially makes all estrogens at least co-carcinogenic, but no one has ever conclusively shown that they are carcinogenic. Estrogens play a role as 'permissive' agents, but that is different from being carcinogenic agents. Something else is needed – one might call this a carcinogenic factor – to trigger off the malignant degeneration, and it may be that presence of hormones is only indispensable at a certain stage.

LAURITZEN: I think that you are right in presuming that every hormone that induces proliferation of a tissue may, under certain circumstances, be co-carcinogenic, but this effect may be dose dependent, and high doses may even inhibit the carcinogenesis or, at a later stage, the development of a carcinoma.

KLOPPER: Professor BAULIEU has pointed to a doubt that many of us are now beginning to entertain: whether there is a special role for estrone among the natural estrogens. It does not matter whether the estrone level in plasma is high or low; as long as there is a proliferative stimulus directed at the endometrium, other factors can cause a carcinoma. This supposition still leaves a special place for estriol. Weight for weight estriol stimulates the endometrium less than the vaginal wall, so there is less of a proliferating stimulus. Moreover, estriol is reasonably competent at excluding other estrogens from combining with endometrial receptors.

THIJSSEN: The mode of administration may be important in obtaining a proliferation. A one-time injection of estriol will not have the same effect as a continuous supply of estriol to the target tissue.

RAURAMO: If you give a high dosage of estriol, which is sometimes needed to suppress climacteric complaints, and divide that dose over the day there is a distinct proliferative effect on the endometrium.

LAURITZEN: Even when there is a proliferation, it will be only slight and inhomogeneous and withdrawal of the estriol does not lead to bleeding. This is also a clinical advantage.

VOIGT: I want to come back on blood levels and what happens at the level of the target organ and make a comparison with prostate hypertrophy. Whereas no differences in blood androgen levels are observed between men with prostate hypertrophy and those without, in the hypertrophic prostatic tissue significantly higher 5-DHT concentrations are found than in normal prostates. This may be explained by an increase in the enzyme activity and/or by the receptor concentration in the hypertrophied organ. One may wonder whether this is a disease-specific change or merely a consequence of ageing.

KELLER: When we look at the blood levels and do not find differences between cancer patients and controls, we should not forget that we have only a momentous picture of the situation at the time the diagnosis is made. We know nothing of what has happened in the years before.

LAURITZEN: It is likely that whether or not a woman will have endometrial cancer is determined early in the first decades of life, before she ever receives estrogen therapy.

SHELLENBERGER: As a toxicologist involved in pre-clinical pharmacology of chemicals and, in this case, of estrogens, I am bothered by what has been said about the facultative effect of estrogens, their being a co-carcinogen rather than a direct carcinogen. How then should estrogens be toxicologically evaluated in the safety evaluations prior to clinical experiments? What animal experiments should we do to gauge the potential risks to humans? The views expressed here may be correct, but it shatters the basis of much toxicological work with hormones.

KLOPPER: It sounds easy, but it is not. The thing to do would be to identify the carcinogen itself because then one would have a model on which to test the co-carcinogens. In any case it seems wrong to concentrate on one molecule. We should look at the whole hormonal picture and include androgens and androgen precursors.

Front. Hormone Res., vol. 5, pp. 101–116 (Karger, Basel 1978)

Uterine Cancer and Estrogen Therapy[1]

THOMAS M. MACK

Cancer Surveillance Program, Department of Pathology, University of Southern California School of Medicine, Los Angeles, Calif.

Introduction

Clinical, epidemiological, and laboratory evidence has long supported a link between female endocrine factors and adenocarcinoma of the endometrium [22]. Based on this evidence, some have long feared that estrogen replacement therapy might increase the risk from this disease. In the past few years several groups, including our own, have purported to test this hypothesis. Each report has described evidence which, in the opinion of its authors, supports a causal hypothesis. These epidemiological studies have been criticized on a number of grounds, and some of these reservations are more sensible than others. My remarks represent an updated version of those I made at an earlier meeting [21]. My purpose is to briefly describe the studies and assess these reservations.

Studies of Endometrial Cancer and Exogenous Estrogens

In the view of an epidemiologist, the most important features of an etiologic study are: the type of study design (follow-up for subsequent disease in a defined cohort versus examination of the historical experiences of cases), the original source of the cohort or cases, the selection criteria for

[1] Supported by a grant from the National Cancer Institute, No. 1 POI CA 17054.

the control subjects to be used as a source of expectation, and the methods available for classifying disease on the one hand and exposure (in this case drug exposure) on the other.

Most of the studies of estrogens and endometrial cancer have been of the case-control type. 30 years ago CORSCADEN and GUSBERG [7] described a series of patients (with no controls), from a New York clinic, 'whose histories are strongly suggestive of an etiologic relationship'. In 1954, JENSEN and OSTERGAARD [18] studied all patients from two Danish hospitals together with a series of postmenopausal controls, otherwise uncharacterized, from other departments of the same hospitals. They concluded that continuous estrogen therapy 'constitutes an important factor in the etiology of endometrial cancer'. 13 years later, DUNN and BRADBURY [9] compared the records of all cases of endometrial cancer (referred to the University of Iowa Hospital) to those of other cases of postmenopausal bleeding (similarly referred). This study is one of the two studies performed to date which found no association between cancer and drug use.

The most recent series of studies began with that by SMITH *et al.* [28] who compared the hospital records of patients referred to several Seattle teaching hospitals for treatment of endometrial cancer with those of patients similarly referred for other gynecologic neoplasms. A strong relationship between estrogen use and endometrial cancer was observed. At the same time, ZIEL and FINKLE [33] compared the records of the incident cases of endometrial cancer occurring among persons enrolled in a prepaid medical insurance scheme with those of a matched set of healthy enrollees. The strong association found was interpreted to be etiologic. Shortly thereafter, MACK *et al.* [20] compared the cases which occurred among the residents of a single retirement community with matched co-residents. Information came from records, from interviews, and from pharmacy records. Again, a strong association, consistent with causality, was found. A year later, GRAY *et al.* [11] abstracted the records of all endometrial cancer cases in a single practice, and of women receiving hysterectomies for benign disease. An association of moderate strength was found. At about the same time, McDONALD *et al.* [23] identified all the cases occurring in one US midwestern county over 30 years, selected matched controls from among co-residents of the same county who had been hospitalized for any reason, and examined all hospital records. This investigation produced evidence of a strong association between prolonged estrogen use and incident endometrial carcinoma.

Preliminary results from three additional studies have been reported at meetings but have not yet found their way into the formal literature. WIGLE and GRACE [32] compared the records of endometrial carcinoma cases from Alberta to those from cases of cancer at other sites. STOLLEY *et al.* [30] compared the records of cases at Johns Hopkins Hospital with those of both gynecological and general hospital patient controls. Both of these studies yielded strong associations between drug and disease. HORWITZ and FEINSTEIN [17] compared the records of cases diagnosed at the Yale-New Haven medical center with the records of two control groups: women with gynecologic cancer at other sites, and women having either dilatation and curettage or hysterectomy for other reasons. Comparison with the first control group yielded a very large relative risk; when the second control group was used, no association was found, as in the other study using the same kind of controls [9]. In addition to these case-control studies, two cohort studies have been performed, both by HOOVER *et al.* [14, 15]. The first of these compared the subsequent endometrial cancer experience of two groups of women (both enrolled in a US national follow-up study), those who did and who did not receive estrogen therapy for breast cancer. An increase in risk was found in those who had taken estrogens. The final cohort study was performed by evaluating the experience of the estrogen users from the same practice providing the cases and controls for one of the case-control studies mentioned above [11]. Subsequent rates of endometrial cancer were compared to various national standards of incidence, each of which produced lower predictions than the observed rates.

As of October 1977, there have been 13 attempts to put to epidemiological test the hypothesis that exogenous estrogens cause endometrial carcinoma. At least four others are in progress (Seattle, Chapel Hill, Ulm and Turku).

Criticisms of the Studies

I classify objections to these studies into three groups: those which are illogical; those which are logical but which cannot reasonably explain the available data; and those which are logical, applicable and upon which attention should be focused.

Let us first examine those logical criticisms of observational studies which, in my opinion, do not offer reasonable explanations of available data.

Real Issues that are no Longer Credible Objections

In a cohort study, if the method of defining the treated and untreated groups is such that they could be expected to experience a different incidence of endometrial cancer under the null hypothesis, i. e. when no biologic association exists, the data are said to be confounded. In case-control studies, the same is true if cases and controls are chosen by criteria which can be expected to result, under the null hypothesis, in a difference in the prevalence of estrogen usage. Needless to say, such a phenomenon can either exaggerate or mask an association, depending on its direction. In order to guard against confounding, it becomes important to recognize determinants of estrogen use (mostly behavioral, such as degree of access to medical care) and known risk factors for endometrial cancer (such as obesity and nulliparity).

The cohort study comparing estrogen users from a single practice with national standards [15] might have produced results confounded by the fact that the women in this practice, on the average, may be more obese, or less parous, or have other attributes conveying high risk independent of any estrogenic effect.

In individual case-control studies the error could be produced in either one or the other direction. The selection of patients with other gynecologic neoplasia as controls [17, 28] could result in an expected estrogen use level based in no small part on women with cancer of the cervix, a disease of the parous and of the poorly educated – characteristics which in the US are associated with a decrease in the intensity of medical care consumption and with, on the average, less easy access to facultative drugs like estrogen [29]. On the other hand, the selection of controls on the basis of having non-neoplastic endometrial disease is problematic for the opposite reason; the women with these commonly prevalent conditions who are admitted to hospitals are likely to have had more than the usual access to physicians, and thus to estrogens. This has been documented when gynecology controls have been compared to general hospital controls [30].

While several of the positive studies can be criticized on the basis of possible confounding, the studies most susceptible to such criticism are the two which produced negative results [9, 17]. Most of the recent positive case-control studies were analyzed in such a way that cases and controls within strata relatively homogeneous for risk were analyzed separately, and high risks for estrogen users were found within such strata. Finally, at least three groups [20, 23, 33] chose controls explicitly to represent the very different specific populations from which their cases appeared. Moreover, in each of these instances, it can be effectively argued that the entire population in question was rather homogeneous in social class and in access to estrogens. Using the prevalence of past estrogen use in the controls as a measure of access, these three studies came up with rather similar relative risks even though that prevalence varied all the way from 50% to 15% to 3%.

Most of the remaining categorical errors which are amenable to correction by study design occur because of one or another kind of bias in classification. It is possible that discrepancies between the morbidity of American women in general and the morbidity of the women in one physician's practice [15] appear because the pathologists serving the single practice use criteria which differ from those used nationally. It is possible that some cases recall drug usage more completely than do controls, that the charts of cases sometimes are abstracted more efficiently than are those of controls, or even that in some settings physicians of women at high risk write more complete notes than the physicians of women at low risk. Any or all of these problems might account for associations of low magnitude in any given study. For example, it seems clear that the Danish study [18] was performed using methods which more efficiently ascertained estrogen history in the cases than in the controls. However, it is probably not reasonable to attribute very strong associations to such errors. Furthermore, several of the investigators took strong measures to guard against such errors, or to detect them in retrospect. The specificity of the findings with respect to other drugs was examined [20, 29, 33]. Abstraction was completely blind in at least one study [33]; in another, three independent measures of drug usage were obtained, including one which was independent of both patient and physician [20]. In this latter study recall bias appeared to work, if anything, in the direction of producing a negative association; 'don't remember' responses (coded as negative) clustered on some of the case forms, probably because of the propensity of some cases to tire quickly.

It is possible in theory that an association between two such phenomena might indicate not causality, but a link between events in the reverse sequence; that the presence of disease might systematically lead to drug usage. Several of the studies avoided that problem by excluding all evidence of usage which occurred or was documented within 6 months or a year of onset.

The question of generalizability has also been raised. Only the study of breast cancer cases [14] would seem to be subject to that criticism; the other studies were of representative cases from definable places and periods.

Unreal Issues

Thus far the criticisms which have been mentioned are important issues in observational studies which do not offer credible explanations for the bulk of the findings to date. There also exists a class of criticisms which essentially raise false issues; they are either not meaningful in themselves or could not explain these findings even in theory.

For example, there has been concern because the endometria of controls in case-control studies were not subjected to pathologic examination and might themselves have contained neoplastic elements. Such misclassification could result in a false-negative study but would not result in a false association; in the extreme it would lead to identical composition of case and control groups.

It has been alleged that the earlier diagnosis in drug users, who presumably receive more intense medical screening, especially for uterine bleeding, might have resulted in a preferential ascertainment of drug usage. Since cases ascertained through *prevalence* screening were not used, since cases in nonusers were no less likely to come to light (albeit at a later age) and since controls were selected by year of birth, there is no reason to suspect that preferential ascertainment occurred.

Moreover, there is empirical evidence to suggest that this effect was not present. No difference between estrogen-associated and nonassociated cases has been noted in the distribution of age at diagnosis. No decrease has been noted in the trend in incidence of invasive disease. Very high relative risks have been noted for women over 70 years of age, thereby 'using up' more cases than would be expected to appear in the remaining life span. The same high relative risk has been shown among cases appearing over the last few years [20, 33] as over the last few decades [23]. High relative risks have been found in patients having symptoms of long duration before diagnosis [20], and in patients presenting with symptoms other than bleeding [11, 30].

Objections have been made to the practice of excluding controls who had had hysterectomies, on the basis that such women are more likely to have received estrogens [29]. Whether they are or not is irrelevant. Such exclusions are necessary since the cases that would have appeared in the absent organs cannot be ascertained and, certainly in most of the more recent studies, no cases which were incidental findings after hysterectomy for other reasons were included in the case rosters. That findings cannot be generalized to women who have received hysterectomies is moot.

Attempts have been made to explain away the cumulative findings on the basis that the exposure promotes, rather than causes endometrial cancer, or that estrogens are co-carcinogenic rather than carcinogenic. Promotion in the laboratory sense cannot be detected at all by the usual epidemiologic studies, because such designs cannot distinguish between accelerated and nonaccelerated progression unless the exposure has just been introduced. If progression is inevitable, a promotor becomes a cause.

However useful it may be as a concept in the context of a specific animal model, co-carcinogenesis is not a distinct general concept from carcinogenesis, either in terms of epidemiologic design or in terms of altering the safety of the environment. As a practical matter *causal* means amenable to use for purposes of prevention.

The findings have also been dismissed for want of a credible physiologic mechanism. This criticism is simply a function of our ignorance. It is true that estrogens have not been shown to be mutagens and are not carcinogens in the usual laboratory sense. Rather than discuss whether or not this statement is more than a terminologic nicety, I will remind you of a similar situation. The hypothesis that diethylstilbestrol given to women during pregnancy causes the appearance of clear cell adenocarcinoma of the vagina in girls delivered to these women also had no precedent in laboratory carcinogenesis. The argument is not weighty in that context and it should not be regarded so in this one.

Finally, there are those who object to these findings simply because they come from studies which are retrospective, or even because they come from studies which are nonrandomized. Both of these objections betray a lack of familiarity with the facts of life in etiologic research. If we restricted our causal inferences in biology to those documentable by the results of randomized studies, we would not only be ignorant but incapable of either advising our patients or predicting the course of their problems. Moreover, as should be apparent from the discussion so far, there is no inherent merit in prospective (cohort) studies vis-à-vis retrospective (case-control) studies. The problems of confounding and bias act in somewhat different ways, but the validity of any given study depends upon the specifics of its design and setting.

Aspects of the Findings which Support an Etiologic Hypothesis

Before completing the discussion of alternative explanations of the information to date which are, in my opinion, inadequate, I should emphasize those findings which positively support an etiologic hypothesis. First on the list is the magnitude of the relative risk estimates. They are very large indeed. Second is the consistency of the studies. Not only have the recent studies been positive but they have been internally and externally consistent in detail. Third, the internal evidences of interaction have been consistent with a biologic relationship. By interaction is meant not only dose-response or duration-response relationships, which have been found in at least six studies, but the tendency for the highest risk from estrogen to occur in women normally at low risk. This may imply a common etiologic pathway between the effects of exogenous estrogens and the 'baseline' etiology of endometrial cancer, or it may only reflect a confirmation of the dose-response relationship (since low risk women, being less obese, may receive a larger dose per kilo from the same regimen). Fourth, the specificity of the association with respect to other drugs in several studies rules out a nonspecific ascertainment bias. Moreover, credibility is enhanced by a lack of specificity in the findings with respect to various classes of steroidal and nonsteroidal estrogens.

Important Issues which Still Remain

The foregoing would suggest that estrogen replacement therapy should cease immediately because of the probability of subsequent cancer morbidity or mortality. In my opinion, objective observers can reasonably temporize by citing either of two loose ends which are just now being tied up. One more serious objection to that conclusion will take longer to resolve.

The first of the two is the possibility that estrogen usage is only a surrogate for another determinant of endometrial cancer. From the evidence to date it seems very unlikely that this factor, if it exists, is one whose effect has already been recognized and quantified. It seems even more unlikely that it is one of the recurrent epidemiological bugaboos like social class or access to medical care. Epidemiologists do, however, have difficulty in distinguishing between the effects of a mode of therapy and the indications for giving it. A woman may be at high risk either because of her treatment, or because of the reasons she was selected for treatment.

In this instance, there is little reason to suspect such a phenomenon *a priori*. The indications for estrogen replacement therapy vary greatly from place to place and from physician to physician. Some women are treated from the first evidences of menopause, *ipso facto* [26], and others are treated only in the presence of persistent recurrent symptoms which interfere with daily functions [24]. Many women with or without such symptoms are not treated because they have not sought treatment. For this reason it is possible in some American communities that women who take estrogens, in the aggregate, differ little in symptomatic profile from women who do not [29]. Furthermore, when recalled symptoms have been studied, they have not explained the association with estrogens [20, 23]. On the other hand, we are profoundly ignorant of the determinants of symptoms of the menopause, and it does seem highly likely that they represent pertinent variations in endogenous endocrine function.

There is other evidence available. If the association with estrogens were really due to an association with underlying symptomatic state, then there would be no reason to expect a correlation between incident rates of endometrial cancer, and estrogen consumption rates in various populations. While such a correlation could have some other basis and is not strong evidence in favor of the hypothesis, the absence of such a correlation does require a bit of explaining if the hypothesis is true.

Only in the US did estrogen usage rates increase in some magnitude sufficiently long ago to permit the expectation of a parallel increase in incidence [29]. And in the US the current prevalence of past hysterectomy in postmenopausal women is sufficiently large and difficult to quantitate that the resultant counter-trend, a decrease in endometrial cancer incidence of sizable magnitude, is a credible possibility [19]. WEISS *et al.* [31] and others [5, 32] have documented a suggestive increase in incidence in North American population-based cancer registry data over recent intervals, and the rates in one wealthy community [20] have been shown to be proportionally high in relation to a pattern of frequent estrogen usage. We are now examining incidence trends in Los Angeles County in the period 1972–1975. While the overall increase in age-adjusted incidence is not striking (fig. 1), the increase in age-specific incidence in the ages following menopause (55–64 and 65–74) (fig. 2) and in women of high social class is more dramatic. Even more striking are the huge differences in age-specific rates, especially among women aged 55–64, by social class (fig. 3). White women resident in census tracts with the highest mean income and education experienced rates which were several times those of poor white

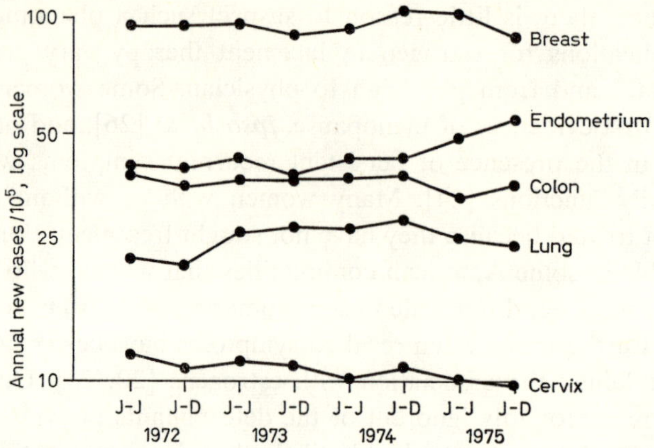

Fig. 1. Age-adjusted incidence of cancer, selected sites. White females. Los Angeles County 1972–1975.

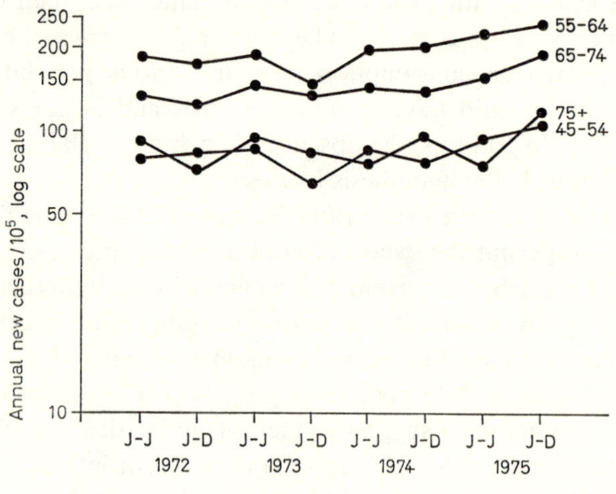

Fig. 2. Incidence of endometrial cancer by age: whites. Los Angeles County 1972–1975.

women, blacks, or women with Spanish surnames. We are currently attempting to measure the patterns of estrogen sales in relation to the same indices of social class.

Others who have examined secular trend in incidence have not found an increase [8]. Today, the best reconciliation of these inconsistent findings is to attribute the failure to a relatively small change in estrogen use over the period or to a compensatory trend in the prevalence of women of

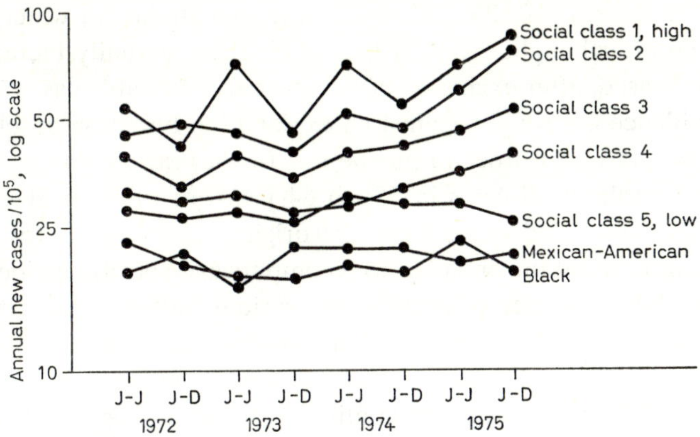

Fig. 3. Age-adjusted incidence of endometrial cancer by race and social class: whites. Los Angeles County 1972–1975.

appropriate age who have lost their uterine corpus [19]. More information is clearly needed, but at the present it is difficult to make a strong case for estrogens as a surrogate for some other factor.

A prudent observer should therefore conclude from available evidence that there is a strong association between estrogen usage, particularly in large doses, and the incidence of a pathologic condition of the uterine endometrium. The next reservation which is not yet fully interred is the one which concerns the nature of the condition. We have long known [12] that prolonged estrogen therapy produces adenomatous hyperplasia, and we must guard against the possibility that some or all of the observed association is due to pathologists conservatively or mistakenly classifying hyperplasia as neoplasia.

Two methods are being used in attempts to evaluate this possibility. In three of the studies [7, 20, 33] the cases of endometrial carcinoma were not reclassified by pathologists using a single set of criteria, but were subdivided into categories of varying degrees of invasiveness or grade, and therefore of varying probability of misclassification. In all three studies the association between estrogens and endometrial cancer held for high as well as low grade lesions, and in two of the three [7, 20] it also held for invasive as well as noninvasive disease.

The second method is to actually submit the slides from the cases studied to panels of pathologists and recompute the relative risks after exclusion of cases using various conservative criteria. This has now been done in three instances [10, 25, 30] and in the two studies in which this was done as an afterthought [10, 25] the relative risk actually increased, rather than decreased, after exclusion of the more questionable cases. Thus the bulk of evidence is strongly against an explanation based on misclassification of hyperplasia. In order to pursue this further in Los Angeles, we have drawn a sample of all cases in 55- to 64-year-old white residents of the county, have requested slides from the original pathologists for purposes of blind interpretation, and plan to recompute age-specific incidences by year and social class after adjustment for various pathological criteria.

The final important issue is one which remains even after making the presumption that estrogens cause endometrial cancer, and to me it is the only truly defensible justification for continuing to use estrogens. It is the quantitative uncertainty which has to be faced in the end by all decisionmakers. How are these presumed costs to be quantitated and weighed against the benefits of estrogen replacement therapy? As a nonpractitioner, I cannot claim to fully answer this question, but I can indicate other questions that must be answered first.

Before doing so it is necessary to emphasize that one question which is *not* pertinent is the crude incidence of disease. Whatever the relative risk for estrogens, the crude incidence is a function of that figure, together with the baseline incidence (in nonusers), and with a figure that can be expected to vary widely, the proportion of women who have been or are users. The quantitative measure of the burden taken on by a woman who becomes a user is best estimated by the attributable risk, calculated by subtracting the baseline risk in nonusers from the risk in users. Note that the proportion of users in the community is irrelevant. If the relative risk is five, then the attributable risk is four times the baseline risk.

If one knows the prevalence of past hysterectomy, the annual incidence of cancer in the population, the proportion of women with intact uteri who are users, and the relative risk, one can estimate the baseline risk and thus the attributable risk. These conditions can be satisfied for three of the study settings [20, 23, 33]. For women aged 55–64 with intact uteri, the estimates of annual incidence per 10^5 in nonusers are (respectively) 80, 52, 75. Estimates of attributable risk for the same three studies are 480, 370, and 300 annual incident cancers per 10^5. Again, these remarkably similar estimates are obtained despite a tenfold difference in the estimated proportion of the population using the drug. For purposes of comparison, the 300–500 annual incident cases would be added to the combined incidence of all cancers (save nonmelanoma skin cancers and cervical cancer *in situ*) for white women of the same age of about 700 annual incident cases per 10^5. Or, looked at another way, a 55-year-old woman taking on this annual attributable risk has taken on a cumulative probability of cancer before reaching age 65 of 3–5 % over the baseline of about 10 %.

But while the quantitative negative impact of estrogens may be sizable, the actual impact on the quality of life may not be. The first unanswered question that should be raised concerns the seriousness of the disease attributable to estrogens. In most of the studies, whatever the degree of association between estrogens and invasive or high grade cancer, there has been a disproportionately high association between estrogens and noninvasive or low grade endometrial carcinoma. Several alternative hypotheses may be offered in explanation. Some of the apparent cancers in this category may actually be hyperplasia. Women receiving estrogens from their physicians may have a lower threshold for scheduling a visit, or may have a shorter wait before being seen, either of which would result, on the average, in an earlier stage. Or there may be a biologic difference between the estrogen-associated cases and the others with respect to rapidity or even the inevitability of progression of disease, even without this being reflected in the usual histologic criteria. Either of these possibilities would also produce an earlier mean stage of diagnosis. Either way, either of these would permit the expectation of an even better improvement in relative survival than that based only upon uniformly early diagnosis. A jump from 80 to 88 % can be predicted on that basis from current US data [3]. In any case, the best guess at present would be that the 55-year-old woman mentioned previously would be taking on only about a 0.3–0.5 % probability of death before age 70 over and above the 20 % she can expect anyway.

The remaining questions that I would call to your attention concern the other medical costs and benefits of those drugs. Epidemiologists have had an extraordinarily difficult time in testing the hypothesis of causality of exogenous estrogens with regard to breast cancer. The information to date has reduced the probability that a large relative risk is involved, but the complexity of other known risk factors for breast cancer, especially late age at menopause, have made it difficult to produce data which are unconfounded by this variable [4], and a relative risk of 2.0 or less certainly cannot be excluded. One recent hint that is worrisome is the higher risk which Hoover *et al.* [16] found accruing to women who had taken higher-dose estrogen tablets in a cohort study which was otherwise too small for firm conclusions. Unfortunately, even if the overall relative risk of 1.3 reported in that study could be certified by some magical process as representing a strictly etiologic association (and at that level the process would have to be magical), the 30% of excess disease would represent about the same probability (0.3%) of unnecessary death before age 70 as does the excess risk from endometrial cancer. This apparent anomaly is explained, of course, by the higher baseline incidence and the higher 5-year mortality of breast cancer. There are undoubtedly other, less important medical side effects of estrogen therapy, such as gallbladder disease, which have not yet been well quantitated [2].

This audience is far more able than I to quantitate the benefits of estrogen replacement therapy in terms of vasomotor symptoms, atrophic vaginitis and osteoporosis. I might only point out that if one ranks those potential consequences of postmenopausal hypoestrogenism which actually threaten life, the most important and obvious single class would be fractures of osteoporotic large bones [1]. Mortality from this cause is probably of the same order of magnitude as that from endometrial cancer. No one has ever tried to estimate the factor by which estrogen replacement diminishes this risk, and we are currently engaged upon such a study. The only other group of serious conditions that might have been considered preventable by estrogens [13], the cardiovascular diseases, cannot now be so considered. Contraceptive estrogens are clearly implicated in cardiovascular events among premenopausal women [27]. Large doses of conjugated estrogens were discontinued from a randomized postmyocardial infarction study because of apparent adverse rather than beneficial effects [6]. While there is now no good evidence that an increase in postmenopausal cardiovascular morbidity or mortality results from estrogen therapy, there is certainly no reason to support the belief that a decrease does.

Summary

Recent studies concerning the relationship of exogenous estrogens and endometrial cancer have been reviewed, and the objections to their conclusions discussed. Even considering the most serious of these reservations, it is difficult to avoid the tentative conclusion that estrogens are causally linked to endometrial cancer, and that the attributable risk is numerically sizable. Whether the medical costs will be ultimately shown to outweigh the benefits of estrogen replacement therapy will depend largely on whether or not a relationship to breast cancer is shown, and on the magnitude of the effect of estrogens in preventing serious complications of ovarian failure. For the time being, estrogenic treatment of menopausal symptoms should be as brief as possible, should use the lowest dose possible, and should probably be given in cyclic fashion, particularly to women with intact uteruses. Every attempt should be made to give patients a quantitative understanding of both potential risks and benefits before they decide for or against estrogen therapy.

References

1 Aitken, J. M.; Hart, D. M., and Lindsay, R.: Oestrogen replacement therapy for prevention of osteoporosis after oophorectomy. Br. med. J. *iii:* 515–518 (1973).

2 Boston Collaborative Drug Surveillance Program: Surgically confirmed gallbladder disease, venous thromboembolism, and breast tumors in relation to postmenopausal estrogen therapy. New Engl. J. Med. *290:* 15–19 (1974).

3 Cancer Patient Survival Report No. 5, USDHEW publication No. (NIH) 77-992 (1976).

4 Casagrande, J.; Gerkins, V.; Henderson, B. E.; Mack, T., and Pike, M. C.: Exogenous estrogens and breast cancer in women with natural menopause. J. natn. Cancer Inst. *56:* 839–841 (1976).

5 Casey, M. J.: Age-specific incidence rates of endometrial cancer. J. Am. med. Ass. *238:* 213 (1977).

6 Coronary Drug Project Research Group: The Coronary Drug Project: Findings leading to discontinuation of the 2.5 mgm/day estrogen group. J. Am. med. Ass. *226:* 652–657 (1973).

7 Corscaden, J. A. and Gusberg, M. D.: The background of cancer of the corpus. Am. J. Obstet. Gynec. *53:* 419–431 (1947).

8 Cramer, D. W.; Cutler, S. J., and Christine, B.: Trends in the incidence of endometrial cancer in the United States. Gynecol. Oncol. *2:* 144–151 (1974).

9 Dunn, L. J. and Bradbury, J. T.: Endocrine factors in endometrium carcinomas. Am. J. Obstet. Gynec. *97:* 465–471 (1967).

10 Gordon, J.; Reagan, J. W.; Finkle, W. D., and Ziel, H. K.: Estrogen and endometrial carcinoma. New Engl. J. Med. *297:* 570–571 (1977).

11 Gray, L. A.; Christopherson, W. M., and Hoover, R. N.: Estrogens and endometrial carcinoma. Obstet. Gynec. *49:* 385–389 (1977).

12 Gusberg, S. B.: Precursors of corpus carcinoma: estrogens and adenomatous hyperplasia. Am. J. Obstet. Gynec. *54:* 905–927 (1947).

13 Higano, N.; Robinson, R. W., and Cohen, W. D.: Increased incidence of cardiovascular disease in castrated women. New Engl. J. Med. *268:* 1123–1125 (1963).

14 Hoover, R.; Fraumeni, J. F.; Everson, R., and Myers, M. H.: Cancer of the uterine corpus after hormonal treatment for breast cancer. Lancet *ii:* 885–887 (1976).

15 Hoover, R. and Gray, L. A.: Menopausal estrogen use and the risk of nonmammary cancer. J. natn. Cancer Inst. (in press).

16 Hoover, R.; Gray, L. A.; Cole, P., and MacMahon, B.: Menopausal estrogens and breast cancer. New Engl. J. Med. *295:* 401–405 (1976).

17 Horwitz, R. I. and Feinstein, A. R.: New methods of sampling and analysis to remove bias in case-control research: No association found for estrogens and

endometrial cancer. Presented at the Society for Epidemiologic Research, 10th Annu. Meet., Seattle 1977.

18 JENSEN, E. I. and OSTERGAARD, E.: Clinical studies concerning the relationship of estrogens to the development of cancer of the corpus uteri. Am. J. Obstet. Gynec. *67:* 1094–1102 (1954).

19 LYON, J. L. and GARDNER, J. W.: The rising frequency of hysterectomy: its effect on uterine cancer rates. Am. J. Epidem. *105:* 439–443 (1977).

20 MACK, T. M.; PIKE, M. C.; HENDERSON, B. E.; PFEFFER, R. I.; GERKINS, V. R.; ARTHUR, M., and BROWN, S. E.: Estrogens and endometrial cancer in a retirement community. New Engl. J. Med. *294:* 1262–1267 (1976).

21 MACK, T. M.: Exogenous estrogens and endometrial carcinoma: studies, criticisms and current status. 2nd Int. Meet. on Endometrial Cancer and Related Subjects, London 1977.

22 MACMAHON, B.: Risk factors for endometrial cancer. Gynecol. Oncol. *2:* 122–129 (1974).

23 McDONALD, T. W.; ANNEGERS, J. F.; O'FALLON, W. M.; DOCKERTY, M. D.; MALKASIAN, G. D., and KURLAND, L. T.: Exogenous estrogen and endometrial carcinoma: case-control and incidence study. Am. J. Obstet. Gynec. *127:* 572–580 (1977).

24 PEARL, M. J. and PLOTZ, E. J.: Management of the climacteric patient. Clin. Obstet. Gynec. *7:* 476–488 (1964).

25 PRENTICE, R.: Endometrial cancer histopathology and survival in relation to exogenous estrogen. Presented at the Society for Epidemiologic Research, 10th Annu. Meet., Seattle 1977.

26 RILEY, G. M.: Endrocrinology of the climacteric. Clin. Obstet. Gynec. *7:* 432–450 (1964).

27 Royal College of General Practitioners: Oral contraceptives and health (Pitman, London 1974).

28 SMITH, D. C.; PRENTICE, R.; THOMPSON, D. J., and HERRMANN, W. L.: Association of exogenous estrogens and endometrial carcinoma. New Engl. J. Med. *293:* 1164–1167 (1975).

29 STADEL, B. V. and WEISS, N.: Characteristics of menopausal women: a survey of King and Pierce counties in Washington 1937–1944. Am. J. Epidem. *102:* 209–216 (1975).

30 STOLLEY, P. D.; ANTONES, C., and ROSENSCHEIN, R.: An investigation of the role of estrogens in endometrial hyperplasia and carcinoma. Presented at the Society for Epidemiological Research, 10th Annu. Meet., Seattle 1977.

31 WEISS, N. S.; SZEKELY, D. R., and AUSTIN, D. F.: Increasing incidence of endometrial cancer in the United States. New Engl. J. Med. *294:* 1259–1262 (1976).

32 WIGLE, D. T. and GRACE, M.: Reproductive Cancer and Hormone Use. Presented at the Society for Epidemiologic Research, 10th Annu. Meet., Seattle 1977.

33 ZIEL, H. K. and FINKLE, W. D.: Increased risk of endometrial carcinoma among users of conjugated estrogens. New Engl. J. Med. *293:* 1167–1170 (1975).

T. M. MACK, MD, Cancer Surveillance Program, Department of Pathology, University of Southern California School of Medicine, *Los Angeles, Calif.* (USA)

Front. Hormone Res., vol. 5, pp. 117–125 (Karger, Basel 1978)

Estrogen Replacement Therapy and Endometrial Carcinoma

L. RAURAMO

Department of Obstetrics and Gynaecology, Turku University Central Hospital, Turku

It is a fact that the long-term continuous administration of large doses of estrogens can induce endometrial cancer in experimental animals. The belief that estrogens may be a causative agent of endometrial carcinoma in the human arose out of this and out of its occurrence in patients with hormone-secreting tumours. This was first mentioned by SCHRÖDER in 1922. When an endometrial cancer is found in young women a Stein-Leventhal syndrome is present in about 20% of cases. Progestagens can cause a regression of hyperplasia and of carcinoma *in situ* of the endometrium. It would seem, therefore, that long-term high doses of estrogens without progestagens may induce endometrial cancer in the human.

There is plenty of evidence in the literature, however, that estrogens as they are normally prescribed at the time of the climacteric do not increase the risk of endometrial cancer. On the other hand, three papers have recently been published in America, which give cause for concern. LAURITZEN [1977] has published an excellent review of the literature on this subject.

SMITH *et al.* [1975] studied 317 patients aged 48 or more with adenocarcinoma of the endometrium and 317 matched controls with cancer of the cervix, ovaries, or vulva. 152 of the patients with adenocarcinoma of the endometrium and only 54 of the controls had used estrogens for at least 6 months. The adjusted relative risk of the exposed women was 7.5 and the unadjusted risk 4.5. ZIEL and FINKLE [1975] reported on a smaller study of the same type with similar findings, and when MACK *et al.* [1976] investigated 63 cases of endometrial cancer and compared them with matched controls from the same community they estimated the re-

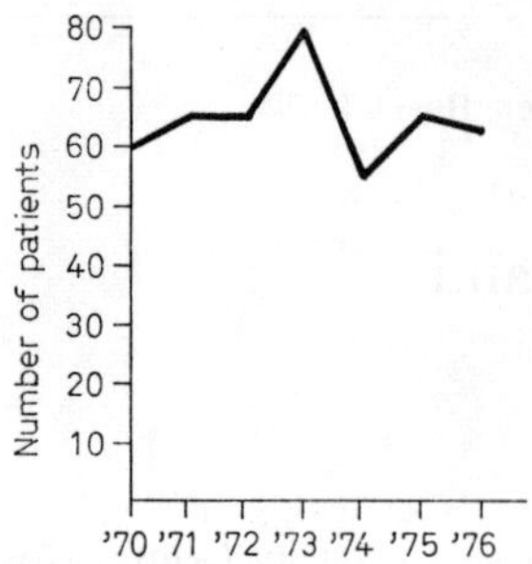

Fig. 1. Incidence of endometrial carcinoma in the years 1970–1976 (total: 453).

lative risk following estrogen therapy to be 8.0, and demonstrated a dose-response relationship.

All three of these studies have been widely criticized, mainly on methodological grounds. Nevertheless, they were very worrying as they indicated a risk which we in Finland had felt not to exist. In order to clarify the situation it was decided to conduct a study in Finland along the lines of that of SMITH *et al.* [1975] but avoiding, where possible, that study's methodological errors.

Material and Methods

It was decided to base our study on interviews of patients in whom endometrial cancer had been diagnosed and treated during the years 1970–1976. The total number of cases detected by us during this 6-year period was 453. The yearly incidence is illustrated in figure 1.

The main reason for our not going back beyond 1970 was that estrogen therapy was less common earlier. It was also felt that relatively few patients could be expected reliably to remember exactly which medicaments they had used so long ago, and it was felt that a retrospective study based only on medical records would be of questionable scientific value.

The 453 patients in whom endometrial cancer had been detected were contacted and asked to come to the Turku clinic for an interview. 317 did so. Of the others, 105 had died, 26 were being controlled by other clinics, and 5 were untraceable.

The control group of 304 women was selected from apparently healthy women who were participating in a mass screening programme for breast cancer. They were matched for age, weight, and social class with the endometrial cancer patients.

The interviews were carried out by Dr. T. SALMI who will, in due course, be publishing full details of the study.

Table I. Hormone therapy used for menstrual disturbances

	Cases %	Controls %
Progesterone	36	48
Progesterone + estrogen	22	25
Estrogen	5	4
Androgen	4	0
Progesterone + progesterone and estrogen	13	16
Androgen + estrogen	2	0
Progesterone and estrogen + androgen	2	0
Progesterone + progesterone and estrogen + androgen	0	3
Not known	16	4

Results

Although the study is not yet completed and no final analysis has been made, a number of interesting facts emerge. Of the endometrial cancer cases, 43% reported that they had received hormone therapy for menstrual disturbances; the average duration was 4 months, and the range 1–27 months. Of the controls, 19% had received hormone therapy for menstrual disturbances; here the average duration was 8 months, and the range 1–25 months. The types of therapy are shown in table I.

It was, perhaps incidentally, interesting to note that a big difference emerged between the two groups when the cancer patients were asked if they had been regularly taking any medicaments at the time their cancer was diagnosed, and the women in the control group were asked the same question for a comparable moment in time. 68% of the cancer patients but only 37% of the controls said that they were. The conditions for which the medicaments were being taken are listed in table II.

More relevant to the present question, however, was the fact that only 29 of the 317 cancer patients (9%) had ever used estrogens, compared to 58 (19%) of the 304 women in the control group.

The mean duration of estrogen use by the cancer patients group was 8 months, ranging from 1 to 46 months, whilst in the control group the mean duration was 15 months, ranging from 1 to 84 months. There was, therefore, considerably less estrogen use in the cancer patients group than in the control group.

Table II. Other diseases and/or medication at the time of the diagnosis of endometrial carcinoma/control

	Cases, %	Controls, %
Hypertension	26	14
Congestive heart failure	22	6
Diabetes	6	1
Analgetics	5	4
Asthma	2	2
Urinary tract infection	2	1
Others	5	9
No regular medication	32	63

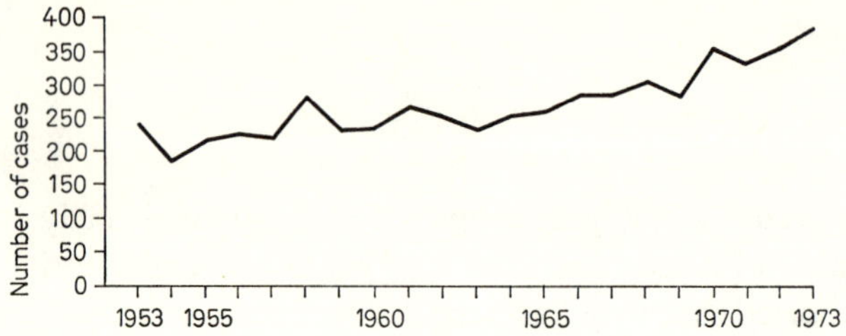

Fig. 2. Cancer of the corpus uteri in Finland: annual numbers of new cases in 1953–1973.

Discussion

From this study it would seem that estrogen replacement therapy is not to be regarded as an aetiological factor for endometrial cancer. The incidence of endometrial cancer has increased in Finland in recent years (fig. 2), just as it has in the United States and many other countries, but this increase seems to have no connection with the increased use of estrogen replacement therapy.

Although the incidence of endometrial cancer has increased, the death rate from this cause has decreased very rapidly. In the United States the incidence of the endometrial cancer rose from 10.3 in 1947 to 20 in 1970. The death rate fell from 9.1 in 1950 to 4.0 in 1970.

In our study there were 9 % estrogen users among the endometrial carcinoma patients and 19 % among the controls. Hence there would seem to be no correlation between the use of estrogens and the occurrence of

Table III. Estrogen use in endometrial carcinoma patients and in controls in three studies

		Cases	Controls
SMITH *et al.*	1975	48 % (152/317)	17 % (54/317)
ZIEL and FINKLE	1975	57 % (54/94)	15 % (29/188)
RAURAMO	1977	9 % (29/317)	19 % (58/304)

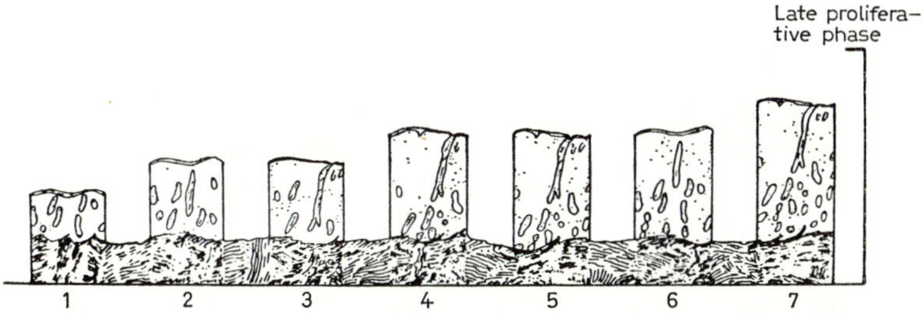

Fig. 3. Endometrium after 6 weeks of estrogen therapy. 1 = Estriol succinate (ES), 2 mg/day; 2 = estradiol valerate (EV), 1 mg/day; 3 = EV, 1 mg/day + ES, 2 mg/day; 4 = EV, 2 mg/day; 5 = EV, 2 mg/day; 6 = EV, 2 mg/day; 7 = EV, 2 mg/day + ES, 2 mg/day.

endometrial cancer. Table III compares this finding with those of two of the quoted reports from the United States.

The explanation for this difference, it is concluded, probably lies in the fact that estrogen replacement therapy is used differently in the United States than it is in Finland.

In the United States estrogens are usually given in relatively large doses often continuously and the preparations most commonly used are ethinyl estradiol and conjugated estrogens. In Finland the only continuously used preparation is estriol succinate (Synapause) in a daily dose of 2 mg, which is too small a dose to produce any proliferation of the endometrium. Estradiol valerate (Progynova) 1 or 2 mg/day is used cyclically. The length of the treatment cycle is 6 weeks in the 1-mg dose patients and 3 weeks in the 2-mg dose patients. The effects of all these drugs on the endometrium of castrated women have been studied, and the results are illustrated in figure 3 [RAURAMO, PUNNONEN and LAUSLAHTI, 1977]. As can be seen, nearly no proliferation occurred with any of them even when the 2-mg dose of Progynova was used twice as long as is usual. It will also be seen that the normal proliferative stage of the endometrium was not reached when estriol succinate and estradiol valerate were used in combination for 6 weeks.

Conclusion

In this study it was found that estrogen use was lower in women with an endometrial carcinoma than in a group of matched controls. The explanation for the discrepancy between this finding and those of recent American reports – in particular that of SMITH *et al.* [1975], the study with which the present one is best compared – may probably be found in differences in the therapeutic approaches in the two countries.

In the United States there is a tendency to use potent estrogens such as ethinyl estradiol and conjugated estrogens, and to use these preparations in high dosages and in a non-cyclical fashion. In Finland, mild estrogens, such as estriol succinate and estradiol valerate are preferred, and these are used in the lowest possible effective dosages. Even when conjugated estrogens are used in Finland, they are given in low dosages and cyclically, if necessary with the addition of a progestagen at the end of each treatment cycle. It must also be mentioned that in Finland, estrogens, whatever their type, are given only under close medical supervision.

Clearly, the combination of these factors makes estrogen replacement therapy a very safe one for the post-menopausal woman.

References

LAURITZEN, C.: Oestrogens and endometrial cancer, a point of view; in GREENBLATT and STUDD The menopause. Clin. Obstet. Gynec. *4:* 145–169 (1977).

MACK, T. M.; PIKE, M. C.; HENDERSON, B. E.; PFEFFER, R. I.; GERKINS, V. R.; ARTHUR, M., and BROWN, S. E.: Estrogens and endometrial cancer in a retirement community. New Engl. J. Med. *294:* 1262–1267 (1976).

RAURAMO, L.; PUNNONEN, R.; LAUSLAHTI, K., and GRÖNROOS, M.: The effect of various doses of oral oestradiolvalerate and oestriolsuccinate on urine calcium/creatinine, serum FSH and endometrium in castrated women. Acta Obstet. Gynecol. Scand. *56:* 363–366 (1977).

SCHRÖDER, R.: Granulosa cell tumors of the ovaries with glandular cystic hyperplasia of the endometrium and beginning carcinoma on this basis. Zentr. f. Gynäkol. *1:* 195 (1922).

SMITH, D. G.; PRENTICE, R.; THOMPSON, D. J., and HERRMANN, W. L.: Association of exogenous estrogen and endometrial carcinoma. New Engl. J. Med. *293:* 1164–1167 (1975).

ZIEL, H. K. and FINKLE, W. D.: Increased risk of endometrial carcinoma among users of conjugated estrogens. New Engl. J. Med. *293:* 1167–1170 (1975).

Prof. L. RAURAMO, Department of Obstetrics and Gynaecology, Turku University Central Hospital, Hämeenkatu 4, *SF-20520 Turku 52* (Finland)

Discussion

LAURITZEN: After the publication of the retrospective studies of SMITH and his co-workers and of ZIEL and FINKLE in which an increase in cancer risk for estrogen users was reported we checked our own patient material.

First of all we proceeded exactly as SMITH *et al.* had done. Our results were much the same. When we looked into the matter, however, we discovered that the registrars had been very conscientious about asking patients with bleedings, mainly, of course, women with endometrial cancer, about their estrogen use, but had rarely asked this information of other patients with cervical, ovarian and vulval cancer. This nullified the usefulness of the data.

We then analysed our material using the technique of DUNN and BRADBURY, which, to my mind, produces far more reliable data [Endocrine factors in endometrial carcinoma. Am. J. Obstet. Gynec. *97:* 465 (1967)]. A curettage was performed on patients with bleedings, and the data grouped according to the endometrial histology. This revealed that estrogens had been taken by 16 % of women with an endometrial hyperplasia, by 11.9 % with an atrophic endometrium, and by only 4 % of those with endometrial cancer. Thus, in our patients, no correlation could be found between estrogen intake and the occurrence of endometrial cancer.

BAULIEU: There are several possible explanations as to why Professor RAURAMO's findings are so different from the American ones. First of all, the dosages used in Finland are lower than in America, and the regimen of treatment is different. Secondly, the estrogens used are different. Estriol is very often used in Finland, but hardly at all in America. Estriol may not be a very strong estrogen, but what is more important is that it is different from estradiol in that it does not bind to the androgen receptors. Whether or not this has any relevance in connection with cancer I do not know, but there certainly is a qualitative difference between this estrogen and the others.

RAURAMO: The study is still going on, and we are not yet in a position to be able to analyse the differences between therapy with estriol and that with estradiol.

HASPELS: A possibly complicating factor in the Finnish study is that data obtained during interviews with patients have been mixed with data taken from the files in the case of patients who died during the course of the study. I wonder if

Professor RAURAMO has, nevertheless, been able to draw any conclusions regarding any differences between the group of patients still alive and those who died.

RAURAMO: Our study started early in 1976 and is based on patients in whom the diagnosis was made after 1969. Few of the patients died in the intervening period, and I do not feel that the fact that data relating to these patients has been taken from their files has adversely affected the study at all. We do not yet have any figures differentiating between the two groups, but it may be that we shall have when we have finished looking at all the material obtained.

STUDD: In no epidemiological studies on this subject have I seen any mention of the mortality rates of endometrial carcinoma. In America the incidence of uterine cancer has doubled in the last 10 years, but at the same time, despite the fact that there were no dramatic changes in therapy, the mortality rate went down from 9.1 to 4. [KEEP, P. A. VAN; GREENBLATT, R. B., and ALBEAUX-FERNET, M.: Consensus on menopause research, p. 62 (MTP Press, Lancaster 1976)]. I understand that SMITH has now reported the 5-year mortality in the population to which his earlier Seattle study referred (lecture). The mortality rate with 'estrogen tumours' was 7 % and with 'non-estrogen tumours' 32 %. This to me, is a very significant difference. What mortality rates have you seen, Dr. MACK?

MACK: The cause-specific mortality has gone down somewhat in the United States, but this is difficult to interpret because of probable errors in the classification. Certainly the stage-specific 5-year survival rate has not changed very much. Like SMITH's, our cases did have a much lower stage frequency distribution. There are three possible explanations for this. (1) There may be cases of misclassified hyperplasia. (2) The second possibility, one that we should not discount, is that these are cancers by any reasonable pathologic definition, but that they do not have as serious an implication, in the long term, as those with which we are familiar. (3) The third possible explanation is that women who are receiving estrogens are followed very closely by their physicians and because of this their cancers are detected at an early stage. In our study the relative risk for the people who have since died was indeed as great as it is for those who have survived. The numbers are too small, however, to conclude that for sure.

The most important observations are those on cases which were blindly screened with only the unquestionable ones being retained. When ZIEL and FINKLE excluded all cases which were questionable in the eyes of even one pathologist the relative risk went up rather than down. The same occurred in SMITH's series.

NORDIN: I have great problems with the word 'causation'. For me something is a cause when it has a predictable and certain effect. The chance of the average woman getting an endometrial cancer is perhaps one in a thousand. If one accepts your figures, a woman on estrogens will have a chance of five in a thousand, but the word 'causation' should not be used for that. At best one could say that estrogens are a risk factor.

Dr. MACK, you showed on a graph that the incidence of cancer went up at the age of 50 to a high peak and then the incidence went suddenly down. I understood, however, that for endometrial cancer in the United States there is a clear rise from the mortality in the 35–39 age group straight up to a maximum in the age group 75–79. In the United Kingdom, where less estrogens are used, the curve flattens off at the time of the menopause [ANTHONY, H. M.: Int. J. Epidemiol. *5:* 231 (1976)].

To me these graphs do not suggest that estrogens *cause* endometrial cancer but that if you do not have estrogens and you have an atrophic endometrium, you cannot get endometrial cancer.

DE WAARD: For the epidemiologist 'causation' is something else. For him, a causal factor is a factor which explains the occurrence of a disorder, even although other factors may well be needed. For an epidemiologist it is not necessary to know the exact mechanism through which a factor works in order to call it a causal one.

NORDIN: That is alright as long as we keep on searching for the real cause.

MACK: I feel quite justified in speaking about 'causality'. In the particular community in which the study reported here was done the attributable risk of endometrial cancer was equal to the entire incidence of breast, colon and cervic cancers together. This is not a negligible addition to the cancer burden. A 'relative risk of eight' means that, in fact 7 out of 8 cases of endometrial cancer in users can be regarded as attributable to estrogen use.

I do not know the data to which Professor NORDIN refers, and so cannot comment on them. There is no question, however, about the correctness of the curves I have shown to you.

NORDIN: It is also odd that the risk of endometrial cancer is so high in your population.

MACK: I also find this extraordinary, but in this particular social class at this particular age the curve is indeed extreme.

LEMON: May hysterectomy, as a very fashionable peri-menopausal event, explain the peak? Once the uterus has gone, there cannot be a second cancer of the endometrium.

MACK: The peak age of hysterectomy is indeed peri-menopausal. In Los Angeles about 35–40 % of all women now 55–64 have had a hysterectomy. In earlier cohorts, such as in women now in their seventies, the figure is much lower. Hysterectomy cannot therefore account for the curve illustrated, because its influence goes in the opposite direction.

Front. Hormone Res., vol. 5, pp. 126–144 (Karger, Basel 1978)

Endometrial Response to Different Estrogens

E. MYHRE

Institute of Pathology, Rikshospitalet, University of Oslo, Oslo

Introduction

Estrogenic hormones are often referred to simply as 'estrogens' as if they were all of the same type having the same effects. This phraseology occurs in official reports as well as in articles in scientific journals. It is, therefore, difficult, if not impossible in many instances, to evaluate the effect of a particular estrogenic hormone.

For the sake of clarity the literally hundreds of different estrogenic substances ought to be divided into three separate main groups: (1) the naturally occurring human estrogens; (2) the conjugated estrogens, which are mixtures of the sodium salts of the sulphate esters of estrone and equine estrogens, and (3) the synthetic estrogens, which may be alkyl-substituted estrogenic steroids or of non-steroidal origin.

Particularly where adverse consequences of treatment with estrogens are concerned, each specific estrogen should be dealt with separately due to their different effects on various parameters, including proliferation of the endometrium. However, reports on specific estrogenic compounds are relatively few compared with the numerous publications dealing solely with 'estrogens'.

Different Estrogens

Synthetic Estrogens

Differences in uterine response to various estrogens have been demonstrated in rats [BRIGGS and BROTHERTON, 1970] (table I) and clinically

Table I. Rank estrogenic potencies by the uterine weight method (after BRIGGS and BROTHERTON, 1970)

Diethylstilbestrol dipropionate	14.37
Estradiol 17-cypionate	11.09
Estradiol benzoate	10.75
Estradiol dipropionate	10.00
Ethinyl estradiol	9.72
Benzestrol	9.28
Diethylstilbestrol	8.76
Estrone	8.59
Estradiol	8.02
Dienestrol	7.73
Promestrol dipropionate	6.83
Diethylstilbestrol dipalmitate	6.60
Sodium estrone sulphate	4.00
Monomestrol	3.26
Hexestrol	2.48
Estriol	2.26
Control	1.00

Table II. Daily substitution doses of oral estrogens in climacteric women and proliferation doses within 14 days (after LAURITZEN, 1975)

Estrogens	Daily substitution doses, mg	Proliferation dose within 14 days, mg
Conjugated estrogens	0.6–1.25 (–3.75)	60–80
Estradiol valerate	1.0–2.0	60–80
Estriol	1.0–5.0	–
Estriol succinate	2.0–6.0 (–8.0)	–
Ethinyl estradiol	0.02–0.04	2.0
Ethinyl estradiol-cyclopentylether (Quinestrol)	0.025–0.05	2.5
Stilbestrol	0.5–2.0	20
Dienestrol	0.5–4.0	30

[LAURITZEN, 1975] (table II). One observes that synthetic estrogens like ethinyl estradiol have a much stronger effect than natural human estrogens. DELFORGE and FERIN [1970] showed that the proliferative effect of *ethinyl estradiol* proved to be 50% stronger that its methyl ether, *mestranol,* in the human endometrium when 100 μg of each substance was used in a sequential regime, but GOLDZIEHER *et al.* [1975], who gave the same substances in doses ranging from 50 to 100 μg/day for two consecutive 21-day cycles, detected no difference between doses or between drugs, indicating that a plateau in endometrial response was reached. A similar plateau was obtained with 50 μg ethinyl estradiol and mestranol in young women with an absence of ovarian function. Mitotic index was lower than during normal menstrual cycles [BROSENS and PIJNENBORG, 1976].

Development of adenocarcinoma of the endometrium has been reported in 4 young women using a preparation containing ethinyl estradiol and dimethisterone, a synthetic progestagen, as long-term sequential oral contraception [LYON, 1975].

The rapidly increasing literature on the pharmacokinetics of the ethinyl estrogens has recently been summarized by HELTON and GOLDZIEHER [1977]; their conclusion at that time was that knowledge of the pharmacokinetic of these compounds was just beginning to accumulate.

Stilbestrol and *diethylstilbestrol* are non-steroidal estrogenic substances with strong proliferative effects on the endometrium. SIROTA and MARINOFF [1975] reported endometrial carcinoma in a 27-year-old woman with Turner's syndrome who had received approximately 15 g diethylstilbestrol over an 8-year period. From the literature they collected seven previous cases of Turner's syndrome who also developed endometrial carcinoma after treatment with large doses of diethylstilbestrol.

Conjugated/Equine Estrogens

The conjugated estrogens contain nearly 50% estrone, while the rest of the preparations contain equine estrogens. To my knowledge little is known about the effects of the equine estrogens on the human endometrium. The effect of estrone is dealt with below.

Natural Human Estrogens

While estradiol is the naturally occurring estrogenic hormone, estrone is also important, though it has less activity than estradiol, and its activity probably depends upon its metabolic conversion into estradiol [FOTHERBY, 1975]. Estriol is regarded as a weak estrogen with hardly any effect upon the endometrium. An excellent survey of the hormonal profiles after the menopause is given by CHAKRAVARTI [1976]. Post-menopausal ovaries make no contribution to estrogen production [BARLOW *et al.*, 1969].

The potencies of the three natural estrogens on the vaginal epithelium and the uterine weight give the relative effect for estradiol/estrone/estriol as 100/10/1 [BRIGGS and BROTHERTON, 1970; LABHART, 1974]. Table III [HASKINS *et al.*, 1968] summarizes the estrogenicity of the three natural

Table III. Estrogenicity relative to estradiol (100%); (after HASKINS *et al.*, 1968)

Investigator and method	Estradiol	Estrone	Estriol
SEALEY and SONDERN			
Immature rat, uterine weight and vaginal cytology (oil)	100	25	10
SZEGO			
Immature and castrated rats, uterine weight (oil)	100	18	256
HUFFMAN and GROLLMAN			
Immature rat, opening of vagina (aqueous)	100	10	33
EVANS, VARNEY and KOCH			
Immature mouse, uterine weight (oil)	100	33	12

estrogens relative to estradiol. It is generally agreed that estriol is less estrogenic than estradiol. One author, however, has indicated to the contrary, that estriol is highly estrogenic when compared to estrone and estradiol. The reason for this is not clear; it may be due to the test conditions.

Estrone circulates in post-menopausal women in amounts of about 40 pg/ml [RADER *et al.*, 1973]. It derives from the conversion of androstenedione from the adrenals to estrone in peripheral tissue [MACDONALD *et al.*, 1968; MATTINGLY and HUANG, 1969; PROCOPÉ, 1969; LONGCOPE, 1971; SAEZ *et al.*, 1972; GRODIN *et al.*, 1973a, b; PROCOPÉ and ADLERCREUTZ, 1973; SIITERI and MACDONALD, 1973; SIITERI *et al.*, 1974]. The aromatization of androstenedione takes place in the fat tissue [NIMROD and RYAN, 1975].

With advancing age the effectiveness of the conversion from androstenedione to estrone increases up to four times the values in young women [HEMSELL *et al.*, 1974]. The normal rate of conversion is about 2%, but this may increase to 11% in post-menopausal women giving an amount of 180 μg estrone/day. If more than 40 μg of estrone is synthesized in this way in post-menopausal women, bleeding may occur [GRODIN *et al.*, 1973a, b]. Levels of estrone amounting to 180 μg/day have been observed in adenomatous hyperplasia and adenocarcinoma uteri [SIITERI, 1975]. From such observations the so-called *estrone theory* has evolved [MACDONALD and SIITERI, 1974], relating the development of endometrial tumours exclusively to the peripheral estrone production. Another argument for this theory is found in the fact that the constitutional stigmata associated with the increased incidence of endometrial tumours are identical to those who give origin to the metabolic mechanism which increases the extraglandular estrone production. The main stigma in this respect is obesity.

Prior to the menopause the ovaries are the most important source of estradiol, but they supply only 50% of the estrone, the remainder coming from the peripheral conversion of androstenedione to estradiol. Androstenedione is secreted by the ovaries and by the adrenals [THIJSSEN, 1976]. After the menopause, practically all estrogenic hormones circulating in the blood derive from the conversion of androgenic hormones into estrogens. The quantity of estrogens resulting from this conversion is greater the higher the body weight of the woman.

All sexual steroid levels with the exception of estrone are significantly lower in post-menopausal women than in young women [VERMEULEN, 1976]. Also androstenedione levels are lower, but the transfer constant is about twice as high in post-menopausal than in pre-menopausal women. This suggests a marked increase in the conversion of androstenedione to estrone with age [SOMMERVILLE *et al.*, 1976]. MARSHALL *et al.* [1976] have confirmed that there is no fall in plasma estrone with age in post-menopausal women.

It is generally accepted that the conversion of androstenedione to estrone is correlated to body weight [MACDONALD *et al.*, 1963]. Circulating estrogen levels are not significantly different in post-menopausal patients with endometrial cancer than in women of similar age *and weight* who do not have the tumour [JUDD *et al.*, 1976]. ROME *et al.* [1977] found that pre-operative estrogen values were higher in post-menopausal women with atypical hyperplasia and adenocarcinoma of the endometrium than in the controls, and further, that the elevated values persisted after oophorectomy.

YEN [1977] pointed to the lack of knowledge of the pharmacokinetic aspects by using exogenous estrogens with different molecular structures. He says that it is possible that the use of estrogens today as far as the choice of drugs, the dosage and so on are concerned is unphysiological. He makes particular mention of the use of Premarin, containing estrone and micronised 17β-estradiol, which leads to a rapid increase in the circulating estrone which reaches a summit after 2–3 h. These daily summits may induce maximal activization of the receptor activity in the target cells.

Estradiol

After oral administration estradiol is rapidly absorbed and results in high plasma concentrations of estrone and of estradiol and of their conjugated 15 min after ingestion. Conjugated estriol reaches a peak after 1½–2 h [LEBECH, 1976].

Experimentally there is an increase in the weight of the uterus and increased endometrial activity as measured by the activity of peroxidase [LYTTLE and DESOMBRE, 1977].

Estriol

According to GOLDFIEN [1977] estrogen therapy is the major cause of post-menopausal bleeding. It is, however, important to know that the histologic picture varies considerably in the material from curettage in post-menopausal women who bleed. The endometrium may display senile/atrophic features, hyperplasia of various kinds and carcinoma, as well as benign uterine or cervical polyps. GOLDFIEN [1977] claims that bleeding occurs in 2–20% of the treated patients depending on the preparation, its dose and on the schedule of administration. LAURITZEN [1973] reported a similar incidence of bleeding in estrogen-treated post-menopausal women, except that estriol displayed less than 1% bleeding on 1.5 mg and 1.3% on 2 mg estriol/day (table IV).

PUCK [1957] maintained, in contrast with earlier views, that estriol is a good preparation when treating the climacteric. It stimulates the cervix, the vagina, and the vulva, and causes no proliferation of the endometrium and, therefore, no bleeding in the post-menopause. The lack of sensitivity of the endometrium to estriol is also emphasized by PEROVIC *et al.* [1975].

Table IV. Frequency of uterine bleedings during administration of various oral estrogens to post-menopausal women (after LAURITZEN, 1973)

Estrogens	Dose mg/day	Percent of women in whom bleeding occurred	Percent of cycles in which bleeding occurred
Conjugated estrogens	1.25	2–12	0.3
	0.6	1–5	0.1
Estradiol valerate	2.0	3–10	0.3
	1.0	2–5	0.1
Estriol	2.0	0.5–1.3	0.03
	1.0	0.2–0.8	0.015
	0.5	0.2–0.8	0.015
Stilbestrol	0.5	10–20	–
Dienestrol	2.0	5–20	–
	0.5	4–14	–
Quinestrol	0.025	5–15	–

WARNECKE [1972] claimed that neither breakthrough bleedings nor withdrawal bleedings occurred when post-menopausal women were treated with estriol. This was also reported by HEUSER and STAEMMLER [1973] who, however, on microscopical examination of the endometrium found slight to medium grade proliferation in most of the 40 post-menopausal patients taking 2–8 mg estriol a day. DEMOL *et al.* [1973] also maintained that endometrial proliferation is not seen with 1–2 mg estriol a day for many months, in contrast to ethinyl estradiol which in doses of 10 μg/day represents a risk of proliferation and bleeding.

Treatment with polyestriol gives estriol as the active component. SJØSTEDT and STRANDH [1971] did microscopical examination in 7 women 2–5 weeks after injection of different doses and found a thin and atrophic endometrium. HASKINS *et al.* [1968], however, have observed withdrawal bleeding after estriol treatment in 3 of 60 patients.

My own experience with estriol treatment of climacteric complaints comprises about 600 patients with a wide age range. Less than half of the patients, namely 259, were post-menopausal. 213 patients (82.2%) were treated with estriol, while the remaining 46 (17.8%) were given estradiol. The regimen with both compounds was: daily treatment for 3 weeks and stop of medication for 3 days. The reasons for this regimen are simply that it is easy to remember, and that the symptoms often reappear after more than 3 days without medication. The dosage varied from 1 mg/day up to 8 mg in a few patients. During the observation time, which until now is short, only between 1 and 3½ years, only 7 patients have experienced slight bleeding. 3 of these were on 3 mg estriol/day, and the other 4 were on 2–4 mg estradiol/day. Curettage of these patients revealed an atrophic endometrium in all cases (fig. 1). The mean age of six of the patients was 51.33 years, while one with previous bilateral oophorectomy was only 30 years old.

Thus, treatment with 3 mg estriol daily resulted in bleeding in 3 of 213 patients which is 1.4%. Treatment with 2–4 mg estradiol daily caused bleeding in 4 of 46 patients, 8.7%. No bleeding occurred when the estriol dose was 2 mg or lower, nor when the estradiol dosage was only 1 mg.

Prolactin has so far not been studied in women in the post-menopause [L'HERMITE and BEN-DAVID, 1976]. There is a clear relationship between the presence of estrogenic hormones and prolactic secretion, but the importance of this is not clear. Estrogen therapy in doses that are effective on the estrogen target organs is insufficient to have a strong influence on prolactin secretion.

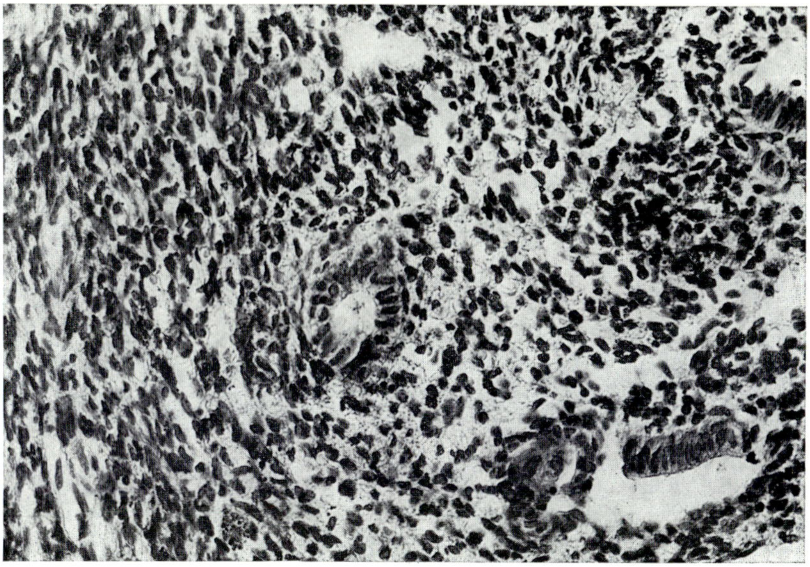

Fig. 1. Atrophic endometrium with fibrous stroma from an estriol-treated postmenopausal woman with slight bleeding.

Endometrial Hyperplasia

Brown *et al.* [1959] related estrogen levels to endometrial morphology. Inactive endometrium was found by 0–10 μg total estrogens secreted per day, proliferation phase by amounts 10–30 μg, and finally cystic glandular hyperplasia by secretion of 25–40 μg/day.

During a normal endometrial cycle the DNA synthesis varies markedly, while RNA synthesis varies less. An important point is that after the addition of *progesterone*, synthesis of both nucleic acids is reduced [Nordqvist, 1970]. It is also known that nor-testosterone derivatives inhibit mitotic activity in general, and also the growth of carcinoma of the corpus [Kaiser, 1959]. A similar effect on mitotic activity is obtained by nor-progesterone [Kaiser, 1969]. Greenblatt and Bryner [1977], using estradiol pellets in the management of menopause, recommended the addition of a progestogen 5–7 days a month to induce withdrawal bleeding and to prevent endometrial hyperplasia.

Experiments have shown that during the secretory phase there is a marked increase in the levels of estradiol 17β-dehydrogenase activity and

Table V. Cystic glandular hyperplasia age distribution (in %)

Age	NOVAK 1924	FALCONER 1947	SCHRØDER 1954	ATTRAMADAL 1970	NILSEN 1977
Under 21	4.5	11	2.6	2.4	0
21–30	16.5	22	4.4	2.0	4
31–40	31.7	29	15.8	11.8	5
41–50	36.3	35	66.8	55.6	58
Over 51	10.6	3	10.4	28.2	29
Number of patients	66	135	3,270	493	96

Table VI. Causes of hyperestrogenism

1	Follicle cysts	6	Hilus cell hyperplasia
2	Polycystic ovaries	7	Arrhenoblastoma
3	Stromal hyperplasia	8	Adrenal cortex hyperplasia
4	Granulosa cell tumours		Adrenal cortex tumours
5	Theca cell tumours	9	Exogenous estrogen

a decrease in the concentration of estradiol receptors [GURPIDE and TSENG, 1976]. The increased dehydrogenase activity also reduced the intracellular concentration of estradiol by an increased conversion to estrone which easily diffuses out of the endometrial cell. Estrone does not bind to receptors in the nucleus as easily as estradiol.

Combination of estrogens with testosterone has been recommended by GLASS [1971] who claims that this mixture has many advantages and no side effects. Using this combination it is also possible to lower the estrogenic dose.

Endometrial hyperplasia may be defined as a continual growth of a tissue which normally undergoes cyclic menstrual degeneration and discharge [SOMMERS, 1970]. During reproductive life hyperplasia of the endometrium usually means that ovulation and normal menstruation have not taken place. Estrogens from various sources (table V) are regarded as the main stimulus for the growth leading to the hyperplasia. Cystic glandular hyperplasia may appear at any age but shows the highest incidence between 41 and 50 years of age (table VI).

There have been several suggestions for the classification of the hyperplastic-neoplastic changes of the endometrium. SHANKLIN [1977] has suggested the terms adenomatous hyperplasia, atypical adenomatous hyperplasia, adenocarcinoma *in situ,* and endometrial carcinoma with or without invasion.

Cystic glandular hyperplasia with atypical changes and adenomatous hyperplasia may in some cases represent early phases in a carcinoma development [HALL, 1957]. So-called pseudo-malignant hyperplasia may be seen in post-menopausal women treated with estrogen [ØSTERGAARD, 1973]. FERIN and THOMAS [1973] reported 43 endometrial biopsies in 34 postmenopausal women aged 50–68 years, all treated 4–14 years. They found cystic glandular hyperplasia in 4 patients and adenomatous hyperplasia in 3 patients. The hyperplastic changes were reversible by the use of progestogens for 6–8 days.

In more than 8.5% of the women with hyperplasia the process ceases naturally, and after curettage there is no recurrence in 70%. The cystic glandular hyperplasia is a relatively remote precursor for carcinoma as less than 6% proceeds to cancer [SOMMERS, 1970]. TAYLOR [1947] and DALLENBACH-HELLWEG [1975] regard the height of the estrogen level to be less important than the duration of its unopposed effect. If estrogen levels are continuously high or moderately high, the condition is called *unopposed estrogenism.*

The cystic glandular hyperplasia may proceed to *adenomatous hyperplasia.* This condition is characterized by the formation of *new glands* in contrast to the cystic glandular hyperplasia where the pre-existing glands show proliferation and cystic dilatation. Adenomatous hyperplasia may regress if the estrogen levels fall. If the levels remain elevated, the adenomatous hyperplasia may show progressive overgrowth which may develop into an adenocarcinoma [DALLENBACH-HELLWEG, 1975]. Before the progression to invasive carcinoma an *adenocarcinoma in situ* may occasionally arise.

NOVAK and RUTLEDGE [1948] claimed that in the majority of cases of adenocarcinoma of the endometrium the malignancy arises in normal, nonhyperplastic endometrium, and that the contrast between the benign and the malignant area is usually clear enough. They also referred to the ability of estrogen to produce pseudo-neoplastic lesions in the uterus and in the breast which has been abundantly established by experimental studies, and they believed that the lesions they described as 'atypical endometrial hyperplasia' belong to this category. BAMFORTH [1956] reported on 81 patients

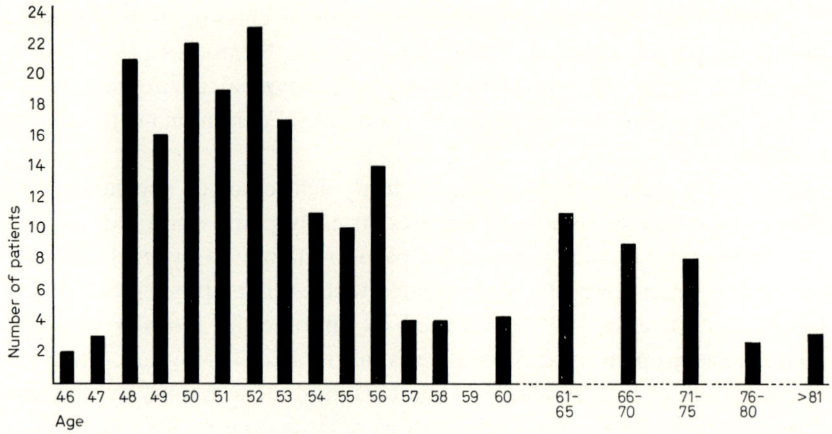

Fig. 2. Cystic glandular hyperplasia. Age distribution.

with carcinoma of the endometrium, and in 17 of these a co-existent cystic hyperplasia was found.

GEIST and SALMON [1941] reported a series of long-continued estrogen administration, concluding that there was no evidence of abnormal proliferation, and that there appeared to be no evidence to justify the fear that carcinoma of the genital tract may result from the therapeutic use of estrogens. The same year, GEIST *et al.* [1941] reported that one case of possible malignancy was found by curettage, but no carcinoma was found in the removed uterus. These authors stressed extreme caution in attributing carcinogenic properties to estrogens in human beings, lest an extremely valuable therapeutic agent be condemned unjustly.

On the other hand, GUSBERG [1947] called attention to a type of endometrial hyperplasia regarded as a cancer precursor, and commented that this condition caused by endogenous or exogenous estrogens may play a role in the development of corpus carcinoma. However, GUSBERG and HALL [1961] reported only 23 cases of uterine cancer in a large series of patients who received estrogen therapy during a 20-year period.

During the last $5^1/_2$ years I have collected 204 cases of cystic glandular hyperplasia in my private pathological laboratory receiving only gynaecological specimens. There were only a few young patients. As I was primarily interested in the use of estrogens in the treatment of the climacteric as a possible cause of the hyperplasia, only patients over 45 years of age were registered (fig. 2). The diagram shows that there was a remarkably high incidence between 48 and 56 years of age. There were also many over 60 years of age. The mean age was 57.4 years. This is remarkably high compared with the age distribution seen in table V.

The questionnaire filled out by the gynaecologists about the use of estrogens by these patients with cystic glandular hyperplasia gave the second surprise. Only 14 patients had received estrogen therapy, making 6.9 % of all the hyperplasia. Those who had been taking estrogens had mostly used synthetic preparations like ethinyl estradiol, a few had used estradiol, and none had used estriol.

Unfortunately, the only information about the patients is that they had strong bleedings. Nothing is known about predisposition like obesity or other conditions. In any case these women are under constant unopposed estrogen influence, which causes cystic glandular hyperplasia.

Discussion

If a considerable number of women in the peri-menopause have an unopposed estrogenism, and if they also have climacteric complaints which are treated with estrogens, it is not difficult to imagine that the treatment may cause further proliferation to atypical hyperplasia and perhaps carcinoma.

Rome *et al.* [1977] investigated the estrogen excretion in post-menopausal women with atypical hyperplasia, adenocarcinoma and mixed adenosquamous carcinoma of the endometrium. They found the pre-operative estrogen values to be higher in the patient groups than in the controls, and that these elevated values persisted after oophorectomy. As a result of this they suggested that an estrogen assay might be useful for screening a post-menopausal population for risk for endometrial cancer. It may be added that an endometrial biopsy would be even more valuable.

These articles, particularly the first two, have been seriously criticized for various reasons which will not be repeated here. One point, however, has not been given sufficient attention: the choice of the estrogenic preparation and the mode of administration, as well as the dosage. In the article of Smith *et al.* [1975] the estrogenic preparation is not even mentioned. They simply stated that estrogen treatment had been given for at least 6 months before the endometrial cancer was diagnosed. In brackets they added that the 'type of estrogen, preparation and dosage were not limiting factors for inclusion on estrogen'. The women Ziel and Finkle [1976] reported upon were given conjugated estrogens, principally sodium estrone sulphate. It is interesting that in a later article [ZIEL and FINKLE, 1976] these authors emphasized that long-term estrogen treatment, *particularly with estrone,* seemed to increase the risk of endometrial cancer. HOOVER *et al.* [1976] used mainly non-steroidal estrogens. In MACK *et al.*'s [1976] paper it is stated that conjugated estrogens were used as well as others, and among these other preparations diethylstilbestrol and ethinyl estradiol occurred with about equal frequency and accounted for the majority of other estrogens specified.

Since these investigations were all performed in the USA, it is highly probable that when the estrogen used is not specified, either Premarin or synthetic estrogens were used.

In SMITH *et al.*'s [1975] paper the endometrial carcinoma was diagnosed after at least 6 months on estrogen treatment. In ZIEL and FINKLE's [1976] paper the duration of conjugated estrogen use before the development of endometrial cancer varied from less than 1 year to several years. HOOVER *et al.*'s [1976] paper is on the development of cancer of the endometrium after hormonal treatment (presumably non-steroidal estrogens) for breast cancer. It is not known how long these patients had been on the estrogen treatment before the development of cancer of the uterine body. Neither is it possible from MACK *et al.*'s [1976] paper to say how long the estrogen treatment lasted before the development of the endometrial cancer.

It is generally accepted that carcinogenesis induced by chemical, physical or hormonal influences takes decades to develop. It is therefore difficult to believe that cancer of the endometrium develops after only 6 months treatment with estrogens unless the endometrium already has a pre-cancerous or cancerous lesion.

The effect of the estrogen therapy may, therefore, have been to provoke a bleeding, which leads to the discovery of an existing cancer. If this is the case, the estrogen treatment may be seen as a very real blessing.

Another point is that if the women treated with estrogens belong to the group of women who already have a rather high estrogen level in their blood, the addition of conjugated estrogens containing estrone as well as non-steroidal estrogens will stimulate any endometrium to grow. If this estrogenism is unopposed, it cannot be denied that proliferation proceeds to various forms of adenomatous hyperplasias which may be a pre-cancerous state.

It is therefore my point that estrogenic substances provoking a strong proliferative effect on the endometrium should not be given to women who might have a high estrogen blood level which already might have provoked excessive endometrial growth.

To minimize the risk of endometrial neoplastic proliferation I advise the following: (1) to give a weak estrogen, like estriol, thus avoiding proliferation of the endometrium; (2) to give estrogens cyclically giving the endometrium a chance to regress during the pause of medication, and (3) to add progestogenic compounds to change the proliferative state into a secretory state. So far secretory states of the endometrium have not been suspected of being pre-cancerous; (4) to give the patient a regular check-up with a curettage if an irregular bleeding occurs.

References

BAMFORTH, J.: Carcinoma of the body of the uterus and its relationship to endometrial hyperplasia. J. Obstet. Gynaec. Br. Commonw. *63:* 415–419 (1956).

BARLOW, J. J.; EMERSON, K., and SAXENA, B. N.: Estradiol production after ovariectomy for carcinoma of the breast. Relevance to the treatment of menopausal women. New Engl. J. Med. *280:* 633–637 (1969).

BRIGGS, M. H. and BROTHERTON, J.: Steroid biochemistry and pharmacology, p. 103 (Academic Press, London 1970).

BROSENS, I. A. and PIJNENBORG, R.: Comparative study of the estrogenic effect of ethinylestradiol and mestranol on the endometrium. Contraception *14:* 679–685 (1976).

BROWN, J. B.; KELLAR, R., and MATTHEW, G. D.: Preliminary observations on urinary oestrogen excretion in certain gynaecological disorders. J. Obstet. Gynaec. Br. Commonw. *66:* 177–211 (1959).

CHAKRAVARTI, S.; COLLINS, W. P.; FORECAST, J. D.; NEWTON, J. R.; ORAM, D. H., and STUDD, J. W. W.: Hormonal profiles after the menopause. Br. med. J. *ii:* 784–787 (1976).

DALLENBACH-HELLWEG, G.: Histopathology of the endometrium; 2nd ed., pp. 94–122 (Springer, Berlin 1975).

DELFORGE, J. P. and FERIN, J.: A histometric study of two estrogens: ethinyl-estradiol and its 3-methyl-ether derivate (mestranol); their comparative effect upon the growth of the human endometrium. Contraception *1:* 57–72 (1970).

DEMOL, R.; HOOGHE, R.; CALLAERTS, J.; HUYGHEBAERT, G., et PROOST, J.: Contribution à la pharmacologie de l'œstriol chez la femme postménopausée. Brux. méd. *53:* 419–431 (1973).

FERIN, J. and THOMAS, K.: Long-term replacement therapy in post-menopausal women. Pituitary gonadotrophin secretion and endometrial studies; in VAN KEEP and LAURITZEN Ageing and estrogens. Front. Hormone Res., vol. 2, pp. 134–142 (Karger, Basel 1973).

FOTHERBY, K.: Oestrogens. Physiological actions; in MAKIN Biochemistry of steroid hormones, pp. 90–104 (Blackwell, Oxford 1975).

GEIST, S. H. and SALMON, U. J.: Are estrogens carcinogenic in the human female? The effect of long-continued estrogen administration upon the uterine and vaginal mucosa of the human female. Am. J. Obstet. Gynec. *41:* 29–36 (1941).

GEIST, S. H.; WALTER, R. I., and SALMON, U. J.: Are estrogens carcinogenic in the human female? II. Atypical endometrial proliferation in a patient treated with estrogens. Am. J. Obstet. Gynec. *42:* 242–248 (1941).

GLASS, S. J.: A safe and effective alternative to estrogen therapy. Am. J. Obstet. Gynec. *127:* 895 (1971).

GOLDFIEN, A.: Estrogen replacement therapy in postmenopausal women. Ration. Drug Ther. *11:* 1–5 (1977).

GOLDZIEHER, J. W.; MAQUEO, M.; CHENAULT, C. B., and WOUTERSZ, T. B.: Comparative studies of the ethinyl estrogens used in oral contraceptives. I. Endometrial response. Am. J. Obstet. Gynec. *122:* 615–618 (1975).

GREENBLATT, R. B. and BRYNER, J. R.: Estradiol pellet implantation in the management of menopause. J. reprod. Med. *18:* 307–316 (1977).

GRODIN, J. M.; SIITERI, P. K., and MACDONALD, P. C.: Source of estrogen production in postmenopausal women. J. clin. Endocr. Metab. *36:* 207–214 (1973a).

GRODIN, J. M.; SIITERI, P. K., and MACDONALD, P. C.: Extraglandular estrogen in the postmenopause; in RYAN and GIBSON Menopause and aging, pp. 15–35. DHEW Publication No. (NIH) 73-319, Washington (1973b).

GURPIDE, E. and TSENG, L.: Estrogens in normal human endometrium; in PASQUALINI Receptors and mechanism of action of steroid hormones, vol. 8, pp. 109–158 (Dekker, New York 1976).

GUSBERG, S. B.: Precursors of corpus carcinoma estrogens and adenomatous hyperplasia. Am. J. Obstet. Gynec. *54:* 905–927 (1947).

GUSBERG, S. B. and HALL, R. E.: Precursors of corpus cancer. III. The appearance of cancer of the endometrium in estrogenically conditioned patients. Obstet. Gynec. *17:* 397–412 (1961).

HALL, K. V.: Irregular hyperplasias of the endometrium. Acta. obstet. gynec. scand. *36:* 306 (1957).

HASKINS, A. L.; MOSZKOWSKI, E. F., and WHITELOCK, V. P.: The estrogenic potential of estriol. A clinical and laboratory re-evaluation. Am. J. Obstet. Gynec. *102:* 665–669 (1968).

HELTON, E. D. and GOLDZIEHER, J. W.: The pharmacokinetics of ethinyl estrogens. Contraception *15:* 255–284 (1977).

HEMSELL, D. L.; GRODIN, J. M.; BRENNER, P. F.; SIITERI, P. K., and MACDONALD, P. C.: Plasma precursors of estrogen. II. Correlation of the extent of conversion of plasma androstenedione to estrone with age. J. clin. Endocr. Metab. *38:* 476–479 (1974).

HEUSER, H. P. and STAEMMLER, H.-J.: Histological investigations into the effect of oestriol succinate on the corpus uteri in postmenopausal women. Arzneimittel-Forsch. *23:* 558–562 (1973).

HOOVER, R.; EVERSON, R.; FRAUMENI, J. F., and MYERS, M. H.: Cancer of the uterine corpus after hormonal treatment for breast cancer. Lancet *i:* 885–887 (1976).

JUDD, H. L.; LUCAS, W. E., and YEN, S. S. C.: Serum 17β-estradiol and estrone levels in postmenopausal women with and without endometrial cancer. J. clin. Endocr. Metab. *43:* 272–278 (1976).

KAISER, R.: Die Wirkung von Gestagenen beim Korpuscarcinom. Arch. Gynaek. *193:* 195–199 (1959).

KAISER, R.: Endokrine Schutzmechanismen gegen Endometrium- und Mammakarzinome. Dt. med. Wschr. *48:* 2467–2473 (1969).

LABHART, A.: Clinical endocrinology. Theory and practice, pp. 576–585 (Springer, Berlin 1974).

LAURITZEN, C.: The management of the pre-menopausal and the post-menopausal patient; in VAN KEEP and LAURITZEN Ageing and estrogens. Front. Hormone Res., vol. 2, pp. 2–21 (Karger, Basel 1973).

LAURITZEN, C.: The female climacteric syndrome: significance, problem, treatment. Acta. obstet. gynec. scand. *51:* suppl., pp. 47–61 (1975).

LEBECH, P. E.: Effects and side-effects of estrogen therapy; in VAN KEEP, GREENBLATT and ALBEAUX-FERNET Consensus on menopause research, pp. 44–47 (MTP, Lancaster 1976).

L'HERMITE, M. and BEN-DAVID, M.: Prolactin and the menopause; in VAN KEEP, GREENBLATT and ALBEAUX-FERNET Consensus on menopause research, pp. 19–23 (MTP, Lancaster 1976).

LONGCOPE, C.: Metabolic clearance and blood production rates of estrogens in postmenopausal women. Am. J. Obstet. Gynec. *111:* 778–781 (1971).

LYON, F. A.: The development of adenocarcinoma of the endometrium in young women receiving long-term sequential oral contraception. Report of four cases. Am. J. Obstet. Gynec. *123:* 299–301 (1975).

LYTTLE, C. R. and DESOMBRE, E. R.: Generality of oestrogen stimulation of peroxidase activity in growth responsive tissues. Nature *268:* 337–339 (1977).

MACDONALD, P. C.; GRODIN, J. M., and SIITERI, P. K.: The utilization of plasma androstenedione for estrone production in women; in GUAL Progress in endocrinology, vol. 184, pp. 770–776 (Excerpta Medica, Amsterdam 1968).

MACDONALD, P. C. and SIITERI, P. K.: The relationship between the extra-glandular production of estrone and the occurrence of endometrial neoplasia. Gynecol. Oncol. *2:* 259–263 (1974).

MACDONALD, P. C.; WIELE, R. L. V., and LIEBERMAN, S.: Precursors of the urinary 11-desoxy-17-ketosteroids of ovarian origin. Am. J. Obstet. Gynec. *86:* 1–10 (1963).

MACK, T. M.; PIKE, M. C.; HENDERSON, B. E.; PFEFFER, R. I.; GERKINS, V. R.; ARTHUR, M., and BROWN, S. E.: Estrogens and endometrial cancer in a retirement community. New Engl. J. Med. *294:* 1262–1267 (1976).

MARSHALL, D. H.; FEARNLEY, M.; HOLMES, H., and NORDIN, B. E. C.: The relation between androstenedione and estrone levels in peri- and post-menopausal women; in VAN KEEP, GREENBLATT and ALBAUX-FERNET Consensus on menopause research, pp. 100–108 (MTP, Lancaster 1976).

MATTINGLY, R. F. and HUANG, W. Y.: Steroidogenesis of the menopausal and postmenopausal ovary. Am. J. Obstet. Gynec. *103:* 679–693 (1969).

NIMROD, A. and RYAN, K. J.: Aromatization of androgens by human abdominal and breast fat tissue. J. clin. Endocr. Metab. *40:* 367–372 (1975).

NORDQVIST, S.: The synthesis of DNA and RNA in normal human endometrium in short-term incubation *in vitro* and its response to oestradiol and progesterone. J. Endocr. *48:* 17–28 (1970).

NOVAK, E. and RUTLEDGE, F.: Atypical endometrial hyperplasia simulating adenocarcinoma. Am. J. Obstet. Gynec. *55:* 46–63 (1948).

PEROVIC, D.; KOPAJTIC, B., and STANKOVIC, T.: Treatment of climacteric complaints with oestriol. Arzneimittel-Forsch. *25:* 962–964 (1975).

PROCOPÉ, B.-J.: Studies on the urinary excretion, biological effects and origin of oestrogens in post-menopausal women. Acta. endocr. *60:* suppl. 135, pp. 80–83 (1969).

PROCOPÉ, B.-J. and ADLERCREUTZ, H.: Oestrogen production in postmenopausal women; in VAN KEEP and LAURITZEN Ageing and estrogens. Front. Hormone Res., vol. 2, pp. 55–73 (Karger, Basel 1973).

PUCK, A. V.: Die Wirkung von Östriol bei Dysmenorrhoe, Entzündungen im weiblichen Genitale, Pruritus und Beschwerden der Klimax. Münch. med. Wschr. *99:* 1505–1507 (1957).

RADER, M. D.; FLICKINGER, G. L.; VILLA, G. O. DE; MIKUTA, J. J., and MIKHAIL, G.: Plasma estrogens in postmenopausal women. Am. J. Obstet. Gynec. *116:* 1069–1073 (1973).

ROME, M.; BROWN, J. B.; MASON, T.; SMITH, M. A.; LAVERTY, C., and FORTUNE, D.: Oestrogen excretion and ovarian pathology in postmenopausal women with atypical hyperplasia, adenocarcinoma and mixed adenosquamous carcinoma of the endometrium. Br. J. Obstet. Gynaec. *84:* 88–97 (1977).

SAEZ, J. M.; MORERA, A. M.; DAZORD, A., and BERTRAND, J.: Adrenal and testicular contribution to plasma oestrogens. J. Endocr. *55:* 41–49 (1972).

SHANKLIN, D. R.: Morphologic findings. J. reprod. Med. *18:* 317–323 (1977).

SIITERI, P. K.: Post-menopausal estrogen production; in VAN KEEP and LAURITZEN Estrogens in the post-menopause. Front. Hormone Res., vol. 3, pp. 40–44 (Karger, Basel 1975).

SIITERI, P. K. and MACDONALD, P. C.: Role of extraglandular estrogen in human endocrinology; in GREEP and ASTWOOD Handbook of physiology. A critical, comprehensive presentation of physiological knowledge and concepts, pp. 615–629 (American Physiological Society, Washington 1973).

SIITERI, P. K.; SCHWARTZ, B. E., and MACDONALD, P. C.: Estrogen receptors and the estrone hypothesis in relation to endometrial and breast cancer. Gynecol. Oncol. *2:* 228–238 (1974).

SIROTA, D. K. and MARINOFF, S. C.: Endometrial carcinoma in Turner's syndrome following prolonged treatment with diethylstilbestrol. Mount Sinai J. Med. *42:* 586–590 (1975).

SJØSTEDT, S. and STRANDH, J.: Effect of polyestriol phosphate on the vaginal cytology and uterine endometrium of postmenopausal women. Acta. obstet. gynec. scand. *50:* 30 (1971).

SMITH, D. C.; PRENTICE, R.; THOMPSON, D. J., and HERRMANN, W. I.: Association of exogenous estrogen and endometrial carcinoma. New Engl. J. Med. *293:* 1164–1167 (1975).

SOMMERS, S. C.: The significance of endometrial hyperplasia and its early diagnosis; in BARBER Gynecological oncology. A comprehensive review and evaluation, pp. 129–140 (Excerpta Medica, Amsterdam 1970).

SOMMERVILLE, I. F.; KITCHIN, Y., and MANSFIELD, M. D.: Endocrine changes at the menopause; in BEARD The menopause. A guide to current research and practice, pp. 41–55 (MTP, Lancaster 1976).

TAYLOR, H. C.: Endocrine factors in the origin of tumors of the uterus; in TROMBLY and PACK Endocrinology of neoplastic diseases, pp. 119–129 (Oxford University Press, New York 1947).

THIJSSEN, J. H. H. and LONGCOPE, C.: Post-menopausal estrogen production; in VAN KEEP, GREENBLATT and ALBEAUX-FERNET Consensus on menopause research, pp. 25–28 (MTP, Lancaster 1976).

VERMEULEN, A.: The hormonal activity of the postmenopausal ovary. J. clin. Endocr. Metab. *42:* 247–253 (1976).

WARNECKE, G.: Die Behandlung des Menopausesyndroms mit Östriol. Z. Ther. *7:* 423–436 (1972).

YEN, S. S. C.: The biology of menopause. J. reprod. Med. *18:* 287–296 (1977).

ZIEL, H. K. and FINKLE, W. D.: Increased risk of endometrial carcinoma among users of conjugated estrogens. New Engl. J. Med. *293:* 1167–1170 (1975).

ZIEL, H. K. and FINKLE, W. D.: Association of estrone with the development of endometrial carcinoma. Am. J. Obstet. Gynec. *124:* 735–740 (1976).

ØSTERGAARD, E.: Malignant and pseudomalignant hyperplasia of the endometrium in postmenopausal women treated with oestrogen; in LEBECH, ULRICH and HELLES Klimakteriet – endokrine, metaboliske og kliniske aspekter, pp. 113–118 (Danmarks Apotekerforening, Danmark 1973).

Prof. Dr. Med. E. MYHRE, Institute of Pathology, Rikshospitalet, University of Oslo, *Oslo 1* (Norway)

Discussion

LAURITZEN: This paper is a very thorough review of the literature on the subject. It is striking that there are so few reports on studies in which the various estrogens, various dose levels and various regimens are compared.

UTIAN: It is good that we ask whether different estrogens have different effects. It is claimed by some that the estrogen receptor is relatively blind; if that were true, the estrogenicity of a substance would be a dose-related response. We have found, however, that this is not the case. When we gave women two different estrogens, both in a proliferative dose, we studied the effect on calcium and on lipids, and found that in this particular group of women one estrogen affected only the calcium plasma levels and the other affected only the plasma lipids. It seems, therefore, that different estrogens do have different pharmaco-dynamic effects, but it is difficult to explain the mechanism of this.

NEVES-E-CASTRO: Professor NORDIN said that our aim should be to transform 'bad' post-menopausal women into 'good' ones, and not try to change them into pre-menopausal women. This remark and the present review prompt me to ask what the profile of the perfect estrogen should be. What are we aiming at? Surely we want more than just relief from vasomotor problems. We are interested in the effects on calcium balance, lipid metabolism, the skin and the psyche, and we want one which will not have adverse effects on the endometrium, coagulation factors or on liver function. It would be a great help in the search for the ideal estrogen for the post-menopause if a profile could be drawn up.

LAURITZEN: You would be completely right if only there would also be just one profile of *the* post-menopausal woman. The 'ideal' estrogen probably does not exist because what is perfect for one woman is unlikely to be so for all. This paper teaches us, for instance, that although the endometrial response to various estrogens is to some extent predictable, there is a huge variation in this response. The same may very well be true of other parameters. This is the disadvantage of the work of the clinician, who has to treat every woman individually; but it is also part of the charm, and keeps us from being prescribing robots.

Front. Hormone Res., vol. 5, pp. 145–154 (Karger, Basel 1978)

Relationships between the Epidemiology and Endocrinology of Mammary Carcinoma

F. de Waard

Institute of Social Medicine, Department of Epidemiology,
University of Utrecht, Utrecht

General Remarks

The efforts of clinical and experimental endocrinologists have produced such a large number of facts and ideas on the causation of breast cancer that it is becoming increasingly difficult to form a clear picture of the present state of affairs.

The epidemiologist should be able to help his colleagues to remain on the right track. In the first place he insists that a possible aetiological factor should fit somewhere into the web of causality spun by the epidemiologists themselves. Where there is no such web, clinicians and experimentalists can talk without restraint. With breast cancer, however, quite a lot is known about its distribution in man. It is for this reason that theories on milk factor viruses and lactation have met with scepticism from epidemiologic circles, for lactation has not shown up as a factor involved in the risk of human breast cancer [21].

Epidemiology provides the concept of attributable risk as well as that of relative risk or risk ratio. Attributable risk as related to the population describes the impact of a particular factor on the total morbidity of the disease under consideration. If a risk factor has a low prevalence among the population amidst other causes, it could – unless it carried a very strong risk ratio – easily go unnoticed when viewing the distribution of that disease as a whole. Conversely, a risk factor of very high prevalence (viz. an endemic factor) can be overlooked because nobody sees a risk in a phenomenon which is so familiar to all.

The attributable risk on a population basis can be expressed as follows:

$$AR\% = \frac{p\,(r-1)}{p\,(r-1)+1},$$

where p = proportion of population exposed, and r = risk ratio or relative risk due to the exposure.

On the well-known question whether the use of oestrogenic drugs could cause endometrial cancer, we had always been told that endometrial cancer had not increased much despite a much greater use of these drugs for relieving postmenopausal complaints. The impact of these drugs did not become visible, however, as long as other possible causes, e. g. obesity (with its consequences of endogenous oestrogen production) were, relatively speaking, much more important than exogenous oestrogens.

As an example: if $p = 0.15$ and $r = 4.0$, the AR% is about 0.31, which means that 31% of endometrial cancers in the population can be attributed to exogenous oestrogens. If, however, $p = 0.05$ with $r = 4.0$, the AR% gets down to 13%. The former assumptions reflect the present situation in a large section of the United States, and several European countries are close followers of American leadership in medicine.

Breast Cancer

Regarding breast cancer, I have attempted to relate our present knowledge on its epidemiology to various facts and ideas contributed by endocrinology.

The role of hormonal drugs has been omitted deliberately. Out of all hormones possibly involved we have chosen prolactin, the androgens and the oestrogens since relevant information on the others (e. g. thyroid hormones) is very scanty indeed.

It is not intended to discuss here all details presented in table I. My remarks are limited to the following:

(1) *Prolactin* – despite its very important role in experimental work with rodents – has not found its proper place in the aetiology of human breast cancer. Extensive studies by Kwa *et al.* [17] have produced largely negative results as far as plasma prolactin levels in high risk families is concerned. If there is indeed a peak at 7 p. m. [18], this hour is inconvenient for practical purposes anyway. If reserpine (which stimulates prolactin release by the pituitary) had proved to be a risk factor, it would fit in very well. However, the initial reports of an association with breast cancer have been followed by several negations [19].

Table I. Relationships between risk factors for breast cancer and hormones

	Prolactin	Androgens	Oestrogens
Ovarian function			castration + [9] anovulation → cont. oestr. [10, 16, 23, 26]
Early menarche	− [17]		+
Late menarche			+
Nulliparity			± [7] E-ratio
Age at 1st pregnancy	− [17]	− [24]	hypothesis [6]
Family history	± [12, 17, 18]	+ [30]	
M. Klinefelter		+	+ ?
Western life-style (nutrition?)	hypothesis [37] + [38]	+ [31] paradox	E-ratio + [5, 8, 11, 20] premenop. − [13] postmenop. + [32]
Body weight and height	− [17]	weight − height ± [34]	weight + [14, 32, 33, 34] height −
Benign (cystic) breast disease		+ [2]	− [25]

Figures within parentheses refer to the list of references.

(2) The role of *androgens,* in spite of tremendous efforts by BULBROOK *et al.* [4], is not at all clear. There is a small bridgehead thanks to the Guernsey study [4], but it is not easy to find an epidemiologic 'corridor' to these results. I suspect that low androgens may enhance the effects of oestrogens at the peripheral level. THIJSSEN *et al.* [28] discovered that androstenediol was able to block oestrogen receptor sites when added in physiological amounts to an *in vitro* system.

(3) There are some leads as to a possible role for *oestrogens,* not only from experimental and clinical but also from epidemiologic evidence.

Dr. LEMON will discuss his ideas on the interaction between various oestrogens and mention the hypothesis by COLE and MACMAHON [6] which is an attempt to fit an oestrogen hypothesis into the epidemiologic pattern. Originally this was mainly to serve the purpose of explaining the risk of a late first pregnancy on breast cancer risk. It has, however, some unexpected relationships with other risk factors, viz. those pertaining to contrasts between the so-called 'Western' and 'Eastern' life-style, a somewhat crude distinction partly related to nutritional habits [8].

We have some confirmation of such a trend from a study of various ethnic groups in Israel [11]. The fact that the ratio between oestrone and oestriol was found to vary not only in premenopausal but also in postmenopausal women indicates that we are dealing with an extraglandular factor in oestrogen metabolism.

It should be made clear that there are as yet no convincing reports of case-control studies which suggest that the ratio between the classic oestrogens is abnormal in patients with breast cancer. Without this, one should be very cautious in calling the so-called oestrogen profile a risk factor.

There are other features of breast cancer which can be linked with oestrogen metabolism. Age-specific incidence curves suggest that from an epidemiologic point of view breast cancer occurring before and after menopause are two different disease entities. Premenopausal breast cancer could have to do with ovarian (dys)function which would be in accordance with the protective effect of early castration [9].

Geographical differences in incidence between Western and non-Western countries are much more profound after than before the menopause. A striking phenomenon is the low incidence in Japan despite its ranking among the industrialised nations. However, with regard to nutrition and nutritional status until recently the Japanese resembled other Asiatic peoples more closely than those from Western Europe and North America.

Having drawn a parallel between endometrial and postmenopausal breast cancer, we studied the relationship between nutritional status (as reflected by weight and height) and hormonal status, in particular signs of oestrogen production in that age group. In a series of cytologic [32, 33] and biochemical studies [34] we repeatedly observed a clear effect of overweight on postmenopausal oestrogens. (Overweight was always expressed as weight adjusted for height.) Recent studies [14] of estrone levels in blood plasma seem to confirm these findings.

In a prospective study on the effect of weight and height on the occurrence of breast cancer we came to our surprise to the conclusion that it was not desirable to adjust weight for height when analysing risk. Actually, tallness itself proved to be a risk factor [35]. The independent effect of height was also found in a number of case-control studies [22, 29]. Recently we confirmed this in a study conducted simultaneously in the Netherlands and in Japan [36] which made it clear that a considerable proportion (> 50%) of the difference in breast cancer incidence between the two countries can be attributed to differences in body size. Here we have an example that not only the relative risk is important but also the proportions of subjects exposed.

Though of great interest from an epidemiologic point of view (height explains the rise of mammary cancer incidence in daughters of Japanese migrants to California and Hawaii) the mechanism of the height effect is still not understood. On the basis of our cytohormonal studies we can exclude an effect on extraovarian oestrogen production.

While exploring the relevant literature it became clear to me that the concept of overweight so familiar in life insurance and public health texts never got a firm foothold in biochemistry and physiology. In fact, in the original papers on metabolic rate, such pioneers as DU BOIS and DU BOIS [1], TEISSIER [27] and KLEIBER [15] had come to the conclusion that the metabolic rate depends on body mass or body size. Throughout the entire spectrum of vertebrate physiology from elephants to mice the metabolic rate can be predicted largely on the basis of body size. Body size is often expressed in terms of body surface area. In physiology this particular choice had much to do with the study of heat loss. From a statistical point of view, however, we are mainly concerned with the fact that surface area is being estimated as a function of height (a dimensional variate) and weight (a variate of mass).

Further research should concentrate on the relation of androgen and oestrogen metabolism to body size and mass. It is now beyond doubt that extraovarian oestrogens are derived from the weak androgen androstenedione. It is conceivable that the distinction between lean body mass (which is positively correlated with height) and fat mass (which is positively correlated with weight) could carry us one stage further in unravelling the metabolic events which promote cancer of the breast.

References

1 DU BOIS, D. and DU BOIS, E. F.: Clinical calorimetry: a formula to estimate the approximate surface area if height and weight be known. Archs intern. Med. *17:* 863–871 (1916).

2 BRENNAN, M. J.; BULBROOK, R. D.; DESHPANDE, N., and WANG, D. Y.: Urinary and plasma androgens in benign breast disease. Lancet *i:* 1076–1079 (1973).

3 BROWNSEY, B.; CAMERON, E. H. D.; GRIFFITHS, K.; GLEAVE, E. N.; FORREST, A. P. M., and CAMPBELL, H.: Plasma DHA sulphate levels in patients with benign

and malignant breast disease. Eur. J. Cancer *8:* 131–137 (1972).

4 BULBROOK, R. D.; HAYWARD, J. L., and SPICER, C. C.: Relation between urinary androgen and corticoid excretion and subsequent breast cancer. Lancet *ii:* 395–398 (1971).

5 BULBROOK, R. D.; SWAIN, M. C.; WANG, D. Y.; HAYWARD, J. L.; KUMAOKA, S.; TAKATANI, O.; ABE, O., and UTSUNOMIYA, J.: Breast cancer in Britain and Japan: plasma oestradiol-17β, oestrone and progesterone, and their urinary metabolites in normal British and Japanese women. Eur. J. Cancer *12:* 725–735 (1976).

6 COLE, P. and MACMAHON, B.: Oestrogen fractions during early reproductive life in the etiology of cancer of the breast. Lancet *i:* 604–606 (1969).

7 COLE, P.; BROWN, J. B., and MACMAHON, B.: Oestrogen profiles of parous and nulliparous women. Lancet *ii:* 596–599 (1976).

8 DICKINSON, L. E.; MACMAHON, B.; COLE, P., and BROWN, J. B.: Estrogen profiles of Oriental and Caucasian women in Hawaii. New Engl. J. Med. *291:* 1211–1213 (1974).

9 FEINLEIB, M.: Breast cancer and artificial menopause, a cohort study. J. natn. Cancer Inst. *41:* 315–329 (1968).

10 GRATTAROLA, R.: The premenstrual endometrial pattern of women with breast cancer. Cancer *17:* 1119–1122 (1964).

11 GROSS, J.; MODAN, B.; BERTINI, B.; SPIRA, O.; DE WAARD, F.; THIJSSEN, J. H. H., and VESTERGAARD, P.: Relationship between steroid excretion patterns and breast cancer incidence in Israeli women of various origins. J. natn. Cancer Inst. *59:* 7–11 (1977).

12 HENDERSON, B. E.; GERKINS, V.; ROSARIO, I. *et al.:* Elevated serum levels of estrogen and prolactin in daughters of patients with breast cancer. New Engl. J. Med. *293:* 790–795 (1975).

13 HILL, P.; WYNDER, E. L.; HELMAN, P.; HICKMAN, R., and RONA, G.: Plasma hormone profiles in populations at different risk for breast cancer. Cancer Res. *36:* 1883–1885 (1976).

14 JUDD, H. L.; LUCAS, W. E., and YEN, S. S. C.: Serum 17β-estradiol and estrone levels in postmenopausal women with and without endometrial cancer. J. clin. Endocr. Metab. *43:* 272–278 (1976).

15 KLEIBER, M.: Body size and metabolic rate. Physiol. Rev. *27:* 511–541 (1947).

16 KODAMA, M.; KODAMA, T.; MIURA, S., and YOSHIDA, M.: Hormonal status of breast cancer. III. Further analysis of ovarian-adrenal dysfunction. J. natn. Cancer Inst. *59:* 49–54 (1977).

17 KWA, H. G.; CLETON, F.; DE JONG-BAKKER, M.; BULBROOK, R. D.; HAYWARD, J. L., and WANG, D. Y.: Plasma prolactin and its relationship to risk factors in breast cancer. Int. J. Cancer *17:* 441–447 (1976).

18 KWA, H. G. and WANG, D. Y.: An abnormal luteal-phase evening peak of plasma prolactin in women with a family history of breast cancer. Int. J. Cancer *20:* 12–14 (1977).

19 MACK, T. M.; HENDERSON, B. E.; GERKINS, V. R.; ARTHUR, M.; BAPTISTA, J., and PIKE, M. C.: Reserpine and breast cancer in a retirement community. New Engl. J. Med. *292:* 1366–1371 (1975).

20 MACMAHON, B.; COLE, P.; BROWN, J. B.; AOKI, K.; LIN, T. M.; MORGAN, R. W.

and Woo, N. C.: Oestrogen profiles of Asian and North American women. Lancet *ii:* 900–902 (1971).

21 MacMahon, B.; Lin, T. M.; Lowe, C. R.; Mirra, A. P.; Ravnihar, B.; Salber, E. J.; Trichopoulos, D.; Valaoras, V. G., and Yuasa, S.: Lactation and cancer of the breast. Bull. Wld Hlth Org. *42:* 185–194 (1970).

22 Mirra, A. P.; Cole, P., and MacMahon, B.: Breast cancer in an area of high parity, Sao Paulo, Brazil. Cancer Res. *31:* 79–83 (1971).

23 Sherman, B. M. and Korenman, S. G.: Inadequate corpus luteum function: a pathophysiological interpretation of human breast cancer epidemiology. Cancer *33:* 1506–1510 (1974).

24 Spicer, C. C.: Androgens and age at first birth in relation to risk of breast cancer; in Host-environment interactions in the etiology of cancer in man. Int. Agency Res. Cancer Scient. Publ. Series No. 7, Lyon, pp. 159–162 (1973).

25 Swain, M. C.; Hayward, J. L., and Bulbrook, R. D.: Plasma oestradiol and progesterone in benign breast disease. Eur. J. Cancer *9:* 553–556 (1973).

26 Swain, M. C.; Bulbrook, R. D., and Hayward, J. L.: Ovulatory failure in a normal population and in patients with breast cancer. J. Obstet. Gynaec. Br. Commonw. *81:* 640–643 (1974).

27 Teissier, G.: La grandeur du métabolisme en fonction de la taille. Etude de biométrie théorique. Annls Physiol. Physiochim. biol. *4:* 1–26 (1927).

28 Thijssen, J. H. H.; Poortman, J., and Schwarz, F.: Androgens in postmenopausal breast cancer: excretion, production and interaction with estrogens. J. Steroid Biochem. *6:* 729–734 (1975).

29 Valaoras, V. G.; MacMahon, B.; Trichopoulos, D., and Polychronopoulou, A.: Lactation and reproductive histories of breast cancer patients in Greater Athens. Int. J. Cancer *4:* 350–363 (1969).

30 Wang, D. Y.; Bulbrook, R. D., and Hayward, J. L.: Urinary and plasma androgens and their relation to familial risk of breast cancer. Eur. J. Cancer *11:* 873–877 (1975).

31 Wang, D. Y.; Hayward, J. L.; Bulbrook, R. D.; Kumaoka, S.; Takatani, O.; Abe, O., and Utsunomiya, J.: Plasma DHA and androsterone sulphates, androstenedione and urinary androgen metabolites in normal British and Japanese women. Eur. J. Cancer *12:* 951–958 (1976).

32 de Waard, F. and Oettle, A. G.: A cytologic survey of postmenopausal estrus in Africa. Cancer *18:* 450–459 (1965).

33 de Waard, F. and Baanders-Van Halewijn, E. A.: Cross-sectional data on estrogenic smears in a postmenopausal population. Acta cytol. *13:* 675–678 (1969).

34 de Waard, F.: Hormonal factors in the epidemiology of breast cancer; in Proc. 4th Int. Congr. Endocrinol. Excerpta med. Int. Congr. Ser. No. 273, pp. 1222–1226 (1973).

35 de Waard, F. and Baanders-Van Halewijn, E. A.: A prospective study on postmenopausal breast cancer in general practice. Int. J. Cancer *14:* 153–160 (1974).

36 de Waard, F.; Cornelis, J. P.; Aoki, K., and Yoshida, M.: Breast cancer incidence according to weight and height in two cities of the Netherlands and in Aichi prefecture, Japan. Cancer *40:* 1269–1275 (1977).

37 WILLIAMS, R. R.: Breast and thyroid cancer and malignant melanoma promoted by alcohol-induced pituitary secretion of prolactin, TSH and MSH. Lancet *i:* 996–999 (1976).

38 WYNDER, E. L.: Epidemiological data about risk groups in breast cancer; in GRUNDMANN Early diagnosis of breast cancer, methods and results. Proceedings of a Symposium at Düsseldorf (Fischer, Stuttgart 1977).

Prof. F. DE WAARD, Instituut voor Sociale Geneeskunde, Rijksuniversiteit Utrecht, Bijlhouwerstraat 6, *Utrecht* (The Netherlands)

Discussion

MACK: Have the age-specific incidence curves to extremes for age at first pregnancy and menarche been constructed for Japan and Rotterdam?

DE WAARD: We have looked at the age at first pregnancy; this was not a confounding factor in studying the effect of weight and height.

LAURITZEN: The effect of the age at the first pregnancy and of the number of pregnancies on cancer incidence is usually presumed to be a hormonal one. I wonder whether it could also be an anatomical one. After all, quite a lot of changes occur in the target tissues: breasts and uterus, during pregnancy.

VAN KEEP: Is it the fact of having been pregnant for some time that is important, or does the pregnancy have to be completed up to delivery to have an influence? If the former, I would be inclined to believe the influence to be a hormonal one.

MACK: It is the age at first full-term pregnancy which is important; the younger the woman the lower the incidence in later life of breast cancer. The abortion has no effect.

LAURITZEN: This is different then from endometrial cancer where patients have statistically less abortions.

MACK: Have you thought about the possibility that height might be the result of some early nutritional phenomenon?

DE WAARD: Yes, we have. The interesting thing is that whilst nutrition apparently affects height at a very early age, its effect on the occurrence of mammary carcinoma comes much later. I am, therefore, inclined to think that the influence of height is associated in one way or another with that of weight.

NEVES-E-CASTRO: The curves for the relation between the incidence of mammary carcinoma and age suggest that we are dealing with two different diseases: one which occurs before the menopause and one which occurs after. One would be inclined to presume that the mammary carcinoma before the menopause is related in one way or another to hormonal factors, whereas that occurring after the menopause seems to be more age dependent. Is there any difference between the two groups in the amount of receptors found in the tissue of tumours?

DE WAARD: Receptor studies in breast cancer tissue have, as far as I know, not yielded much epidemiological information. Receptors are not found in all cancers which occur before the menopause. We think the ovary may have something to do with premenopausal breast cancer, but we do not know the exact mechanisms.

NEVES-E-CASTRO: Has the age at menopause an effect on the incidence of mammary carcinoma? If so, could this partly explain the difference you found between the Netherlands and Japan?

DE WAARD: The median age at menopause in the Netherlands is slightly over 51 years of age. I do not know the exact age in Japan. It is interesting to note that the hook in the curve has changed slightly in the Netherlands. It used to be, as in many other countries, around the age of 50 but now comes slightly later, at about 55. We do not know whether there is a secular trend in the age at menopause as older studies into the age at menopause are not very reliable.

NEVES-E-CASTRO: The relationship between weight, the estrogenicity of the vaginal smear, and the eventual occurrence of mammary carcinoma, is indeed very striking. Would it be reasonable, therefore, to regard a reduction in weight as a preventive measure in connection with mammary carcinoma, and perhaps also for endometrial carcinoma?

DE WAARD: We believe that weight reduction may well be a primary preventive measure, but this is not easy to demonstrate. We are just beginning a prospective study with periodic screening comparing consecutive mammograms, and perhaps we shall then be able to demonstrate an association between the growth rate of the cancer and body weight. We expect that an association will only be seen in those cases in which estrogen receptors are found; in these women it may indeed be wise to bring the estrogen levels down by bringing down the body weight.

NORDIN: Are you implying that estrogen therapy may also increase the risk of breast cancer?

DE WAARD: If it was not for the height factor, I would think that there is a strong analogy between estrogens and the incidence of endometrial and breast cancers, but as long as the height factor is not explained, I would not want to say that.

LAURITZEN: You mentioned that the incidence of breast cancer in Japanese women in Hawaii has increased. Is this correlated with a higher intake of estrogens in the post-menopause?

DE WAARD: I do not have any data regarding the estrogen intake of these women. I must say, however, that I do not regard estrogens as *the* factor in the epidemiology of breast cancer.

LAURITZEN: The fact that height and weight seem to have such a clear influence makes me wonder if the growth hormone, somatostatin, or perhaps prolactin, play a role in the occurrence of the disease.

DE WAARD: I am open to any hypothesis which can explain the effect of height. We really do not know how to explain this at the present time. The most likely explanation I can see is that the body's total size may influence in one way or another certain metabolic processes.

LEMON: How many women were there in the group 'tall post-menopausal women with breast cancer'? There cannot have been many as there are not that many tall Japanese women.

DE WAARD: There were 322 women with breast cancer in the Japanese of our case-control study; 272 were between 35 and 69 years of age and 171 of those were taller than 150 cm [Cancer *40:* 1269–1275 (1977)].

LEMON: In Detroit, WOLFE, a leading mammographer, has found that about 20% of patients with a duct proliferation of the peri-menopause develop breast cancer in a period of about 5 years. In another 20% the duct proliferation resolves and the mammogram returns to normal. I think that the work of WOLFE should be watched. It may be that it will provide us with a means of recognizing women at risk.

STUDD: This is a very important point. Mammography is obviously the best way of studying breast tissue, as one cannot frequently do biopsies of the breast the way one can of the endometrium.

DE WAARD: There are still a lot of questions that need to be answered about mammary carcinoma. In an attempt to answer some of them we have taken nine samples from a large number of women, about 20,000, who are being screened regularly. These have been frozen, and will be used later, in conjunction with mammographic studies, to see whether differences exist between women who develop mammary carcinoma and those who do not.

MACK: Our group has also looked at WOLFE's work, and has found that, particularly in younger women, the presence of the duct proliferation indicates an increased mammary carcinoma risk of about twenty-fold [WILKINSON, E.; CLOPTON, C.; GORDONSON, J.; GREEN, R.; HILL, A., and PIKE, M. C.: J. natn. Cancer Inst. *59:* 1397–1400 (1977)]. Only one mammogram is needed for classification purposes. Repeated mammography adds nothing to the information first gained and may in fact contribute to the development of a breast cancer later in life.

BAULIEU: Personally, I think, that from a hormonal standpoint, there is a danger in regarding carcinoma of the breast and carcinoma of the endometrium as completely comparable diseases. One does not find a sizeable amount of estrogen receptors in about one third of the cases of breast cancer, but I have not yet seen one single endometrial cancer where there were less estrogen receptors per cell unit than the highest concentrations observed in the pre-menopausal endometrium. The estrogen receptor concentration in the endometrium is always high, whatever the state of differentiation of the endometrial cancer. The endometrial cell is more dependent than the mammary cell on estrogens, and I am inclined to believe that the same is true for tumours.

Front. Hormone Res., vol. 5, pp. 155–173 (Karger, Basel 1978)

Clinical and Experimental Aspects of the Anti-Mammary Carcinogenic Activity of Estriol[1]

H. M. LEMON

University of Nebraska Medical Center, Department of Medicine,
Section of Oncology, Omaha, Nebr.

Introduction

Estrogen therapy for menopausal disorders has wide clinical popularity, but is fraught with increasing concern for the possible hazards of promotion of breast or endometrial carcinomas in those so treated. BYRD and others have followed up women receiving conjugated estrogen therapy after salpingo-oophorectomy and hysterectomy, finding identical or increased incidence rates of carcinoma of the female breast compared to age- and race-matched control populations in the USA with intact ovaries, suggesting that the use of hormonal replacement therapy has nullified the potential reduction of breast cancer incidence rates to be expected after castration [4, 12]. The increasing incidence rate of endometrial carcinoma in recent decades, especially on the west coast of the United States, now exceeding carcinoma of the uterine cervix has raised serious questions about the safety of estrogen therapy after the menopause [9].

Estriol is excreted throughout life in both sexes, and following menopause remains one of the principal identifiable constituents of urinary estrogen metabolites. During pregnancy, the feto-placental unit produces about 200–1,000 μg/kg/24 h of estriol, which is rapidly excreted. Our interest in the possible anti-carcinogenic activity of estriol developed during our investigations of fractional urinary estrogen excretion in Caucasian women with mammary carcinoma, two-thirds of whom excreted estriol/estrone + estradiol-17β excretion quotients (estriol quotients) below 0.7, compared to only 20% of controls, with or without various benign diseases [19].

[1] The partial support of these studies by G. W. CARNRICK C., 65 Horse Hill Road, Cedar Knolls, N. J., is gratefully acknowledged.

Castration or menopause, androgen or estrogen therapy and operative stress increased this quotient [20, 21].

Estriol possesses several rather unique properties:

(a) Unlike estrone, estradiol-17β, or diethylstilbestrol, estriol administered in high chronic doses (150–200 μg/kg/24 h or more) has seldom been reported to have a significant carcinogenic activity in rodents.

(b) Estriol co-administration with estrone or estradiol to spayed rodents leads to both acute (3 day) and chronic (2 weeks) reduction of uterine growth. Estriol administered in ratios as low as 0.13 (estriol/estrone), 0.5 (estriol/estrone + estradiol-17β) or 1.0 (estriol/estradiol-17β) resulted in notable arrest of uterine growth in castrates [11]. The reduction of estradiol-induced uterotropic activity by estriol was dose dependent, over a 15-fold variation in estriol dosage. These observations were confirmed and extended other C6 and C16 hydroxylated derivatives [13]. C2-hydroxy-estrogens bind to uterine receptor proteins far in excess of their uterotropic potency [29]. Both 2-methoxy-estrone and 2-hydroxy-estradiol-17β have been reported as having activity equal to estriol in reducing estradiol stimulation of uterine growth [31].

(c) Estriol enhances phagocytic activity, unlike estradiol or estrone, as a part of the immune response in mice [31].

The discovery of uterine estrogen receptor proteins a decade after the initial observations of HISAW *et al.* [11] developed a biochemical basis for the competition between estrogenic hormones to gain access to uterine or mammary intracellular binding sites. A spectrum of relative binding activities of the major estrogens to cytoplasmic receptor proteins has been described [3, 17]. The brief (2–3 h) duration of binding of the estriol-receptor complex after transfer to nuclear chromatin appears to account for the minimal uterotropic activity of estriol given as a single small dose (0.1–10 μg) in contrast to estradiol-receptor complexes, 20% of which persist for over 6 h in the nucleus [1, 15]. WEBSTER and HAMILTON [36] have described reduced activation of uterine RNA polymerases II and III by estriol, compared to estradiol-17β which activates all three of these polymerases. However, such observations fail to explain the marked impairment of uterine growth *in vivo,* observed with combined estriol and estradiol administration in spayed rodents.

MACMAHON *et al.* [28], COLE *et al.* [5], and many others have reported biochemical epidemiological data relating early age of first pregnancy as a protective factor against subsequent breast carcinoma in Caucasian women and have found that parous women after completion of pregnancy excrete higher estriol quotients than nulliparous. A negative correlation between urinary estriol quotients in the non-pregnant pre-menopause woman, or in post-menopause women, and their risk of subsequent development of breast cancer has also been reported from many countries [7, 10, 22, 28].

Experimental Anti-Mammary Carcinogenic Activity of Estriol

Experiments substantiating an anti-mammary carcinogenic function for estriol were initiated in our laboratory in 1968. The effect of intermittent measured doses of estrogens, androgens, progestins and glucocorticoids starting at 50–55 days of age in intact female Sprague-Dawley rats, upon the incidence of induced mammary carcinoma appearing during the subsequent natural life span of these rats has been observed (table I) [23, 24]. 48 h after subcutaneous implantation of the hormone as a known percentage of a crystalline NaCl pellet, mammary carcinogenesis was induced by oral administration of 20 mg 7,12-dimethylbenz(a)anthracene (DMBA), 50–70 mg of procarbazine (PC), or 350 rad whole body radiation delivered by a cobalt teletherapy unit [2]. The hormone-containing pellets were re-implanted every 60 days subcutaneously as long as each rat lived. The rats were maintained in suspended individual cages, supplied with tap water and Wayne Lab Blox feed *ad lib.* They were examined for tumors every 1–2 weeks, and were only sacrificed when a growing tumor was demonstrated on repeated examinations approaching or exceeding 1.0 cm in diameter. All biopsied or spontaneously dying animals were necropsied, with measurement of uterine dry and wet weights in those killed by cervical dislocation. All tumor biopsies were interpreted without knowledge of the treatment of the host, following a standard classification [30].

Appropriate control groups of untreated rats, steroid hormone-treated, carcinogen-treated, in addition to carcinogen- and steroid-treated rats from a single supplier were tested and the results expressed in two ways:

(a) Incidence rates of animals with one (or multiple) breast carcinoma in each group of rats, have been reported with inter-group differences in rates analyzed by Chi square test using the Yates correction for small numbers [23]. The median longevity of each group was also tested for comparable duration of observation.

(b) More recently, a single expression of results has been developed and will be reported here, based upon the total number of breast cancers observed in each tested group, per thousand rat-days of observation.

The tabulation of induced breast and other cancers is now based on observations from over 2,600 rats observed for their natural life span.

Recently, RNA metabolism of liver, uterus and mammary fat pads has been investigated in 55- to 60-day-old female Sprague-Dawley rats not receiving any carcinogen, after castration, or after implantation of 600-μg pellets of estriol subcutaneously. Our objectives have been to compare the early effects of castration and estriol therapy upon metabolism of intact and transformed uterine and breast epithelium. Initially the method of KIRBY [16] was used for phenol extraction of RNA from freshly excised and frozen (–70 °C) tissue samples. RNA was quantitated by its ultraviolet absorption spectrum at 2,650 Å units in a Beckman DU spectrophotometer. 24 h prior to sacrifice each rat received 1.6×10^7 d. p. m. of ($^{6\text{-}14}$C) orotic acid hydrate intravenously in the tail vein, and the specific activity of incorporated orotic acid per milligram. RNA was determined in each sample [35]. Pooled tissues were used from 2 rats in each case, which included specimens from all four mammary fat pads.

Estriol 2.5–15 mg daily has also been administered to 30 post-menopausal women with carcinoma of the breast, for periods up to three plus years, with observations of the vaginal cytology maturation index, and serum cholesterol, triglyceride fraction and total lipids, as well as clinical evaluation of their menopausal symptoms and breast cancer metastases.

Observations

Experimental Anti-Mammary Carcinogenesis

A total of 24 natural and synthetic hormones have been evaluated for their anti-mammary co-carcinogenic activity after subcutaneous implantation as pellets every 2 months in our rat test system, using DMBA as the carcinogen in most instances (table I). Estrone and estriol thus far have been the only two implanted hormones in these intact female rats to significantly reduce breast carcinoma incidence ($p < 0.01$) after chemical induction of carcinogenesis, although statistically insignificant reduction of breast carcinoma incidence occurred with the use of several other estrogens, especially in PC-treated rats (fig. 1). Estriol was most highly inhibitory of breast carcinoma development for the lifetime of the female rat, of any of the natural estrogens or their metabolites. Since rat tissues possess 16α-hydroxylase activity, producing estriol from estrone or estradiol [25], it is likely that the observed reduction of breast carcinoma incidence rates

Table I. Hormones tested for anti-mammary carcinogenic activity in intact female Sprague-Dawley rats (600–1,200 μg dose every 2 months)

Estradiol-17β	Dehydroepiandrosterone
Estradiol-17α	Testosterone
16,17-Epiestriol	Progesterone
17-Epiestriol	Corticosterone
16-Epiestriol	
16-Keto-estradiol-17β	
2-OH-estradiol-17β	
D-Equilenin	
DL-Equilenin	
6-Dehydro-estrone	
2-OH-estriol	
Hexestrol	
Diethylstilbestrol	
Estrone-3-SO_4	
Estriol-3-methyl ether	
17α-Ethinyl-estradiol-17β	
17α-Ethinyl-estriol	
17α-Ethinyl-estriol cyclopentyl ether[1]	
Estriol[2]	
Estrone[3]	

[1] Reduction of breast carcinoma incidence compared to DMBA- and PC-treated controls significant, $p < 0.001$ [23].
[2] Reduction of breast carcinoma incidence compared to PC-treated controls significant, $p < 0.005$ [23].
[3] Experiments in progress.

may in part be due to metabolism of estriol precursors such as estrone to estriol; however, 16α-hydroxy-estrone was an outstanding exception, being wholly inactive in reducing breast cancer incidence in one experiment. A second experiment in progress has confirmed its inactivity. Neither 17α-ethinyl-estradiol-17β, hexestrol nor diethystilbestrol in equimolar doses compared to estriol resulted in any reduction of breast carcinoma incidence; none of these compounds can be metabolized to estriol [24]. Methylation and 2-hydroxylation of estriol led to loss of all activity; estriol epimers and 16-keto-estradiol-17β were also inactive.

Estriol reduction of induced mammary carcinogenesis was maximal using pellets containing 600–1,300 μg of steroid, implanted every 2 months (fig. 2). As little as 150 μg of estriol implanted every 2 months significantly reduced breast carcinogenesis in PC-treated female rats, but was ineffective

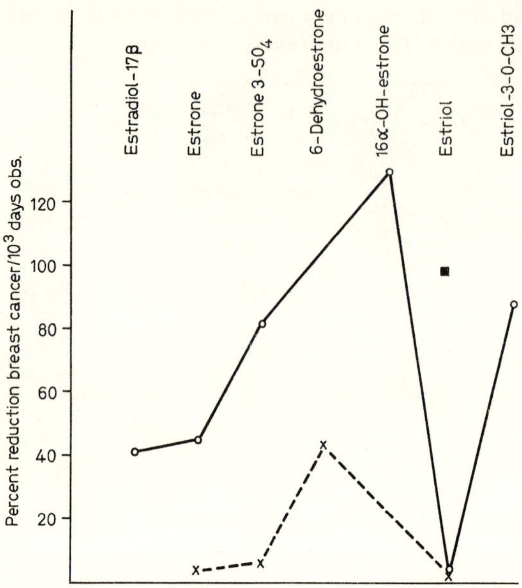

Fig. 1. Relative activity of estrogen metabolites in reducing experimental mammary cancer incidence. The percent reduction of breast cancer incidence on the ordinate is derived from:

$$\frac{\text{Total No. ca. breast in hormone + carcinogen-treated/10}^3\text{ days observation}}{\text{Total No. ca. breast in carcinogen-treated controls/10}^3\text{ days observation}}.$$

The hormonal precursors of estriol are serially listed along the abscissa. ○ = DMBA-treated rats. × = Procarbazine-treated rats. Only estriol has been tested in irradiated rats (■).

in DMBA-treated rats in whom 600 μg estriol was the minimum effective dose for rats weighing less than 250 g 4 months after DMBA treatment. Higher estriol doses of 1,300 μg implanted every 2 months were necessary for prevention of most breast carcinomas in more obese rats exceeding 250 g mean body weight. The implantation of 600 μg of estriol monthly led to a lesser reduction of breast carcinoma incidence, although the numbers of rats tested is such that the differences observed are not significant. Serial re-implantation of estriol every 2 months beginning 48 h before carcinogen exposure was required for maximal breast cancer suppression. A similar degree of inhibition of mammary carcinogenesis by estriol was observed using either of the two chemical carcinogens to induce cancers, but no reduction of gamma ray induced breast cancers has been observed after estriol administration (fig. 1). The inhibition of mammary carcino-

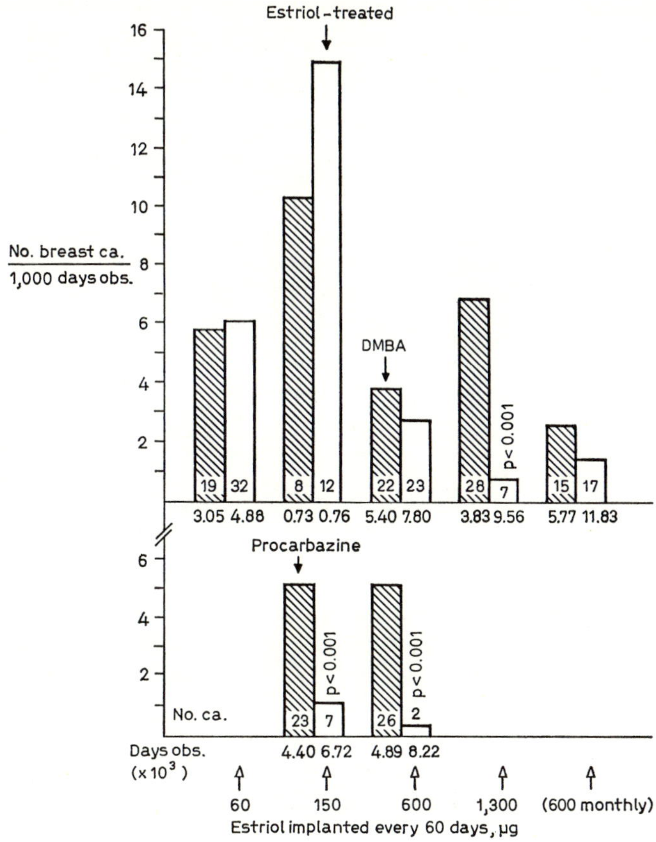

Fig. 2. Estriol dose-dependent reduction of chemically induced breast carcinoma, after DMBA (above) or procarbazine (below) induced cell transformation. Total number breast carcinoma/10^3 days observation (ordinate) plotted against increasing estriol dosage over the range of 60–1,300 μg every 2 months (abscissa). Carcinogen-treated controls for each dose cross-hatched; estriol + carcinogen-treated open bars. P values determined by Chi square applied to incidence of breast cancer-bearing rats in each experiment [23].

genesis by estriol administration by intact females was equivalent to the reduction in mammary carcinogenesis observed when castration preceded DMBA administration (fig. 3). Control studies failed to reveal any spontaneous mammary cancers in our strain of rats during their natural life span, and only a single mammary cancer developed in a steroid-treated control rat after 26 months, 1 of 5 breeders implanted with 1,200 μg es-

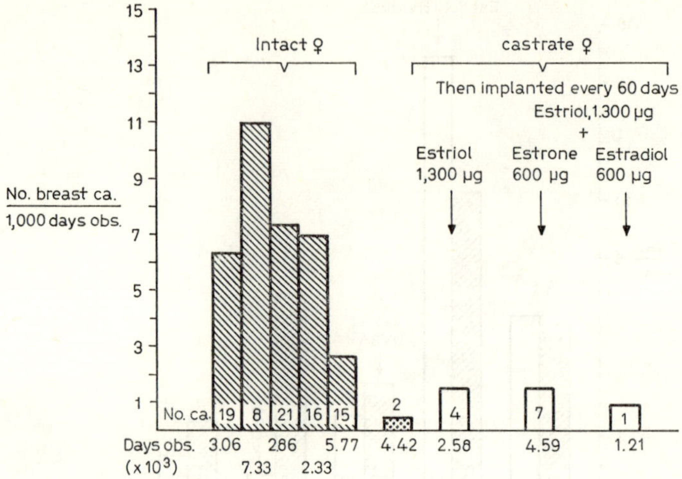

Fig. 3. Reduction of 7,12-dimethylbenzanthracene-induced breast carcinogenesis by prior castration. Total number breast carcinoma/10^3 days observation (ordinate) as in figure 2, comparing somewhat earlier contemporary DMBA-treated carcinogen control results in intact females, with previously castrated females fed DMBA, with and without subsequent hormone replacement.

triol every 2 months. Estriol 1,300 μg alone, or in combination with simultaneously administered estrone or estradiol-17β, administered every 2 months resulted in only a minimal increase in mammary carcinogenesis in DMBA-fed castrate females (fig. 3).

Anti-Uterotropic Activity of Estriol Pellets in Intact Female Rats

Although 100% estriol pellets were reported to take as long as 90–100 days for absorption from subcutaneous depots in rats [8], addition of ^{14}C-estriol to the 600 μg non-radioactive estriol dispersed throughout the 5–7 mg NaCl pellets indicated much briefer absorption (fig. 4). Urinary excretion of lipid soluble ^{14}C indicated that estriol absorption was complete by 2–3 weeks. Excretion data for conjugated ^{14}C material could not be obtained owing to severe quenching of fluorescence. Since the weight of our rats usually plateaued between 250 and 300 g, we have revised our maximum absorption rate for 600 μg estriol pellets during the first 7–10 days, to be about 180 to 240 μg/kg/24 h, and decreasing thereafter.

In spite of the rather brief absorption of estriol from the pellets, a comparison of wet and dry uterine weights obtained at the time of rat sac-

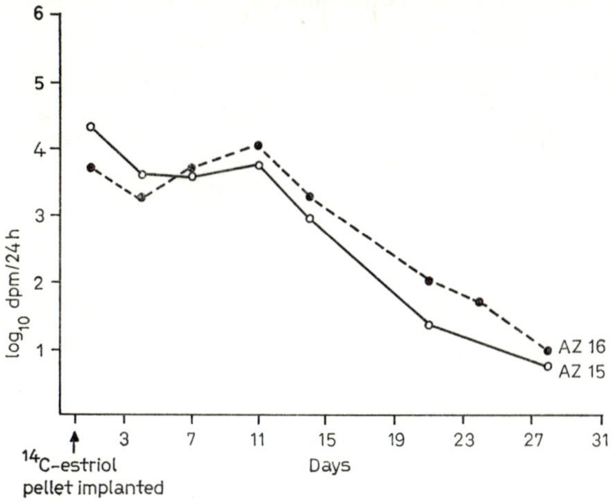

Fig. 4. Excretion of ether soluble ^{14}C following implantation of ^{14}C-estriol pellets subcutaneously. Log_{10} dpm/24 h (ordinate) vs. time in days (abscissa).

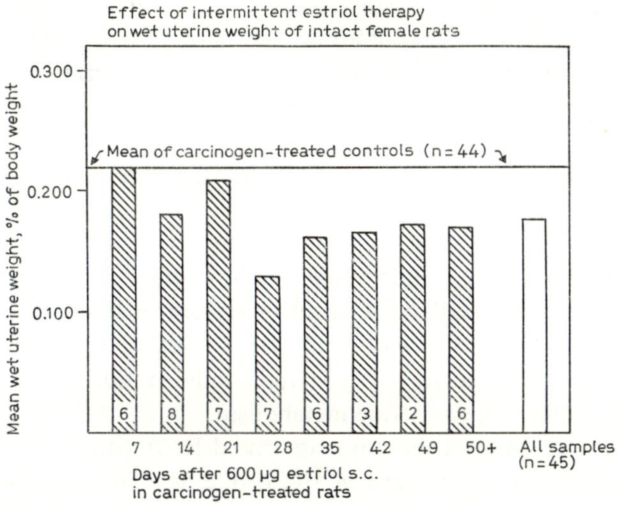

Fig. 5. Wet uterine weights in estriol-treated intact females at the time of sacrifice for biopsy of neoplasms compared to those in carcinogen-treated controls of same experiment. Uterine weight as percent of total body weight (ordinate) is compared up to 60 days after pellet implantation (abscissa), with the mean percent uterine weight of sacrificed carcinogen-treated control rats. Number of samples in each time period at base of each bar.

Table II. Effect of castration and estriol implantation on RNA metabolism (experiment IV)

	Mg RNA/g wet weight and ^{14}C dpm/mg Mg RNA					
	liver		uterus		fat pads	
Procedure	mg	dpm	mg	dpm	mg	dpm
Control	0.0918	1,580	0.6530	47	0.1253	702
Castrate						
1 week	0.1986	2,145	0.3650	1,140	0.2330	2,052
2 weeks	0.1178	1,740	0.4130	1,659	0.1876	5,886
600 μg estriol 1 week after implant	0.1689	1,285	0.1632	699	0.4420	434
600 μg estriol 2 weeks after implant	0.1766	832	0.4230	676	0.2670	884

rifice for biopsy of tumors in carcinogen control and estriol- plus carcinogen-treated rats, indicated a sustained reduction of uterine weights in the latter group (as percent of total body weight) (fig. 5). Other 'impeded' estrogen metabolites such as 16-epiestriol also reduced uterine weights, but synthetic estrogen such as hexestrol did not.

Comparison of Uterine RNA Metabolism after Castration or Estriol Implantation

1 and 2 weeks after castration, or the implantation of 600 μg of estriol, uterine RNA concentration per unit tissue weight decreased by 25–50 %. This was accompanied by a 10- to 15-fold increment in the specific activity of incorporated ^{14}C orotic acid (table II), suggesting that the early changes in uterine weight occurring after estriol administration were similar to castration in the effects upon uterine metabolism.

Both castration and estriol therapy augmented RNA concentration in mammary fat pads, with an increase in ^{14}C orotic acid incorporation at 24 h in the castrate material, which was not seen after estriol implantation.

Estriol Administration in Human Post-Menopausal Metastatic Breast Carcinoma

These studies will be reported elsewhere, but the principal observations may be summarized as they pertain to the post-menopausal state.

5 mg of estriol administered daily induced a 10- to 15-fold increase in the percentage of cornified cells in the vaginal smear, and after 4–7 months uterine bleeding in 6 patients. There were no trends noted in body weight or blood pressures. Mild increases in plasma triglyceride and cholesterol fraction were noted in some patients. About one-third of treated patients showed objective evidence of partial remission or arrest of their metastases, chiefly noted in soft tissue local recurrences, while one-quarter with osseous metastases noted early exacerbation of pain symptoms indicating increased growth activity. Control of menopausal symptoms in two younger women in their mid-40s post-oophorectomy required 10–15 mg of estriol daily for 80–90% relief of vascular symptoms. In over 13 patient years of treatment, no adverse changes have been noted in the contralateral breast or uterus in these women receiving 2.5–15 mg estriol daily without interruption.

Discussion

Both procarbazine and 7,12-dimethylbenzanthracene are equally highly potent mammary carcinogens, whose mechanisms of cell transformation probably differ considerably. It is unlikely that estriol in some way interferes with the transforming activity of both of these chemical carcinogens upon breast tissue to prevent cancer. Estriol probably functions as a mammary anti-co-carcinogen, by impeding ovarian estradiol stimulation of breast duct and alveolar cell replication, preventing the 10^{30} cell generations necessary for a palpable 1.0 cm tumor to become apparent. Its action would therefore be comparable to castration, and indeed our observations confirm this. The early effects of estriol implantation, or castration upon uterine RNA metabolism also support this concept. However, estriol is the only one of the C2, C6 or C16 'impeded' estrogens thus far tested capable of reducing mammary carcinogenesis, suggesting that anti-uterotropic activity is not predictive of cancer-protective activity. Daily administration of much larger doses of synthetic estradiol antagonists (1,000 μg/kg/24 h) have also been reported to reduce DMBA-induced mammary cancer incidence comparable to castration [33]. In these investigations, a perturbation of estrogen receptor function was postulated, because of 90% sequestration of estrogen receptors in the nucleus, after treatment with the estradiol antagonist.

The ineffectiveness of estriol therapy upon gamma irradiation induced mammary carcinogenesis has no explanation at present, unless the dosage used was too low. 10 mg estradiol implantation prior to whole body radiation reduced breast cancer incidence by 50 % [2].

In Caucasian pre-menopausal women there is considerable variation of estriol excretion rate in the urine with about 20 % excreting reduced estriol quotients below 0.7 [22]. After one or more normal full-term pregnancies, estriol excretion appears significantly increased in younger premenopausal non-pregnant women, compared to nulliparous controls, especially in the luteal phase [5]. This suggests that the protective effect of early pregnancy upon subsequent breast cancer risk may not be entirely due to hyperestriolemia during pregnancy. Criticism of the protective role of estriol against breast cancer has centered about the very low plasma concentrations of free estriol in non-pregnant women, compared to estrone or estradiol [26]. However, estriol is not bound to the testosterone estradiol binding globulin of plasma, nor to serum albumin, to the same extent as estrone or estradiol, so that a higher fraction of circulating estriol gains access to extracellular fluid, which is the critical point for competition between estrogens for intracellular access. A recent investigation of the relation between urinary estriol quotients and the endogenous production rate of estriol and other estrogens has shown little correlation between these two parameters [27], but confirmed a 4-fold variation of endogenous estriol biosynthesis, with 20 % of subjects in the lowest category. Confirmation of this study is necessary using other possibly more reliable methods.

6 mg daily of estriol administered orally to healthy pre-menopause women elevates the plasma-free estriol to concentrations seen in the first two trimesters of pregnancy and has had little effect upon menstrual function or ovarian hormone output [34].

Our investigations of estriol therapy of post-menopausal women agree with other investigations as to the hormonal potency of estriol in altering some of the objective signs of ovarian insufficiency, and the rather insignificant effects upon plasma cholesterol and triglyceride fraction [18]. The therapeutic activity of continuous daily estriol therapy upon metastatic breast cancer appears comparable thus far to that of diethylstilbestrol in our series, and cannot be compared in any way to its potential value as a prophylactic for breast or endometrial carcinogenesis. To mention only one major difference, the normal breast and benign tumors of the adult female breast seldom have identifiable titers of free estrogen receptor proteins, comparable to those found in many breast carcinomas. Estriol pretreatment of patients with breast cancer has not been found to reduce the uptake of tritiated estradiol in biopsied metastases or primary tumors [6]. Most of the measured uptake in this investigation, however, was to non-specific protein binding by albumin and other tissue components; estrogen receptor proteins were not isolated from cytoplasm or nuclei and tested for their specific activity.

Because of its safety in high plasma concentrations both in mother and fetus, estriol remains an attractive candidate for a controlled prospective clinical trial, testing its potential anti-mammary carcinogenic activity in the young pre-menopausal women with low urinary estriol excretion quotients. It also appears to be the safest natural product for use in treatment of the menopausal syndrome, even though its therapeutic effect upon vascular manifestations of the menopause require higher doses than most other estrogens including conjugated equine estrogen therapy [18]. Experimental osteoporosis in rats has been prevented by feeding 100–120 μg estriol/kg/24 h [14]. Since estrone has approximately equal uterine receptor cytosol binding activity as estriol, the chances of significant reduction of abnormal estrone-induced endometrial hyperplasia and carcinoma by chronic low dose estriol therapy appears equally likely as breast carcinoma prophylaxis [32]. Our investigations of the anti-mammary carcinogenic activity of estriol suggest thus far that intermittent estriol administration may be the most effective means of blocking the co-carcinogenic activity of ovarian steroids upon the breast epithelial tissues, especially in Caucasian women with reduced endogenous estriol biosynthesis, or elevated estrone or estradiol biosynthesis with a normal estriol production rate.

Summary

Intermittent implantation of 600–1,300 μg estriol subcutaneously beginning 48 h before oral administration of 7,12-dimethylbenz(a)anthracene or procarbazine prevents development of 80–90 % of carcinomas of the breast occurring during the natural life span of the intact female Sprague-Dawley rat. Some estriol precursors were less inhibitory of breast cancer development among 23 other estrogens and androgens, progestins and glucocorticoids tested. More frequent or lower estriol doses than 100–200μg/kg/24 h every 2 months were less inhibitory of breast carcinogenesis. No other types of neoplasms were reduced in incidence by estriol implants, which also reduced uterine weights by 20–25 %. Intermittent substitution of estriol for estrone or estradiol in the nuclear receptor complexes of target cells probably accounts for these observations, which resemble the effect of castration in reducing breast cancer incidence. Human studies indicate excellent tolerance for oral estriol doses of 10–200 μg/kg/24 h, which may correct subnormal estriol/estrone + estradiol urinary quotients associated with elevated risk of breast carcinogenesis in epidemiologic investigations.

References

1 ANDERSON, J. N.; PECK, E. J., and CLARK, J. H.: Estrogen-induced uterine responses and growth; relationship to receptor estrogen binding by uterine nuclei. Endocrinology *96:* 160–167 (1975).

2 BOGDEN, A. E.; COBB, W. R.; CATE, C. C.; KNO, E. Y. H., and ALEX, S. A.: Inhibition of x-radiation induced mammary tumorigenesis by 17β-estradiol. Abstract 22. Proc. Am. Ass. Cancer Res. (1975).

3 BRECHER, P. I. and WOTIZ, H. H.: Stereospecificity of the uterine nuclear hormone receptors. Proc. Soc. exp. Biol. Med. *128:* 470–473 (1968).

4 BYRD, B. F.; BURCH, J. C., and VAUGHN, W. K.: Significance of post-operative estrogen therapy on the occurrence and clinical course of cancer. Ann. Surg. *177:* 626–631 (1973).

5 COLE, P.; BROWN, J. B., and MACMAHON, B.: Oestrogen profiles of parous and nulliparous women. Lancet *ii:* 596–598 (1976).

6 DESHPANDE, N.; CARSON, P., and HORNER, J.: Oestriol in human breast tumors. J. Steroid Biochem. *7:* 11–14 (1976).

7 DICKINSON, L.; MACMAHON, B.; COLE, P., and BROWN, J. B.: Estrogen profiles of Oriental and Caucasian women in Hawaii. New Engl. J. Med. *291:* 1211–1213 (1974).

8 FORBES, T. R.: Further observations of the relative absorption rates of pellets of various crystalline compounds implanted subcutaneously in rats. Endocrinology *30:* 761–769 (1942).

9 GORDON, J.; REAGAN, J. W.; FINKLE, W. D., and ZIEL, H. K.: Estrogen and endometrial carcinoma; an independent pathology review supporting original risk estimate. New Engl. J. Med. *297:* 570–571 (1977).

10 GROSS, J.; MODAN, B.; BERTINI, B.; SPIRA, O.; DEWAARD, F.; THIJSSEN, J. H. H., and VESTERGAARD, P.: Relationship between steroid excretion pattern and breast cancer incidence in Israeli women of various origins. J. natn. Cancer Inst. *59:* 7–11 (1977).

11 HISAW, F. L.; VELARDO, J. T., and GOOLSBY, C. M.: Interaction of estrogens on uterine growth. J. clin. endocr. Metab. *14:* 1134–1143 (1954).

12 HOOVER, R.; GRAY, L. A.; COLE, P., and MACMAHON, B.: Menopausal estrogens and breast cancer. New Engl. J. Med. *295:* 401–405 (1976).

13 HUGGINS, C. and JENSEN, E. V.: The depression of estrone induced uterine growth by phenolic estrogens with oxygenated functions at position 6 or 16; the impeded estrogens. J. exp. Med. *102:* 335–346 (1955).

14 IGARASHI, M.; HAYASHI, Y., and KARUBE, S.: Effect of estriol on bone collagen of rats with experimental osteoporosis. Endocr. Jap. *21:* 387–391 (1974).

15 JULIANO, J. V. and STANCEL, G. M.: Estrogen receptors in the rat uterus retention of hormone receptor complexes. Biochemistry *15:* 916–920 (1976).

16 KIRBY, K. S.: Biochem. Prep. *6:* 79 (1958).

17 Korenman, S. G.: Comparative binding affinitive of estrogens and its relation to estrogenic potency. Steroids *13:* 163–177 (1969).

18 Lauritzen, C.: The management of the pre-menopausal and the post-menopausal patient. Ageing and estrogens. Front. Hormone Res., vol. 2, pp. 2–21 (Karger, Basel 1973).

19 Lemon, H. M.; Wotiz, H. H.; Parsons, L., and Mozden, P. J.: Reduced estriol excretion in patients with breast cancer prior to endocrine therapy. J. Am. med. Ass. *196:* 1128–1136 (1966).

20 Lemon, H. M.: Endocrine influences on mammary cancer formation. A critique. Cancer *23:* 781–790 (1969).

21 Lemon, H. M.: Abnormal estrogen metabolism and tissue estrogen receptor proteins in breast cancer. Cancer *25:* 423–435 (1970).

22 Lemon, H. M.: Review: Genetic predisposition to carcinoma of the breast. Multiple human genotypes for estrogen 16-alpha-hydroxylase activity in Caucasians. J. surg. Oncol. *3:* 255–273 (1972).

23 Lemon, H. M.: Estriol prevention of mammary carcinoma induced by 7,12-dimethylbenz(a)anthracene and procarbazine. Cancer Res. *35:* 1341–1353 (1975).

24 Lelon, H. M.: Estrogen structure-activity correlations for prevention of chemically induced mammary carcinoma. Abstract No. 234. Proc. Endocrine Soc. (1976).

25 Lemon, H. M.: Unpublished data.

26 Lipsett, M. B.: Oestrogen profiles and breast cancer. Lancet *ii:* 1378 (1971).

27 Longcope, C. and Pratt, J. H.: Blood production rates of estrogens in women with differing ratios of urinary estrogen conjugates (in press, 1977).

28 MacMahon, B.; Cole, P., and Brown, J. B.: Etiology of human breast cancer; a review. J. natn. Cancer Inst. *50:* 21–42 (1973).

29 Martucci, C. and Fishman, J.: Uterus estrogen receptor binding of catechol estrogens and of estetrol (1,3,5 (10) estratriene 3,15 alpha, 16 alpha and 17β-tetrol). Steroids *27:* 325–333 (1976).

30 Murad, T. M. and Von Haam, E.: Studies of mammary carcinoma induced by 7,12-dimethylbenz(a)anthracene administration. Cancer Res. *32:* 1404–1415 (1972).

31 Nicol, T.; Vernon-Roberts, B., and Quantock, D. C.: Oestrogenic and anti-oestrogenic effects of oestriol, 16-epi-oestriol, 2-methoxy-estrone and 2-hydroxy-oestradiol 17-beta on the reticulo-endothelial system and on the reproductive tract. J. Endocr. *35:* 119–120 (1966).

32 Siiteri, P. K.; Schwarz, B. E., and MacDonald, P. C.: Estrogen receptors and the estrone hypothesis in relation to endometrial and breast cancer. Gynecol. Oncol. *2:* 228–235 (1974).

33 Tsai, T. L. S. and Katzenellenbogen, B. S.: Antagonism of development and growth of 7,12-dimethylbenz(a)anthracene-induced rat mammary tumors by the anti-estrogen U23,469, and effects on estrogen and progesterone receptors. Cancer Res. *37:* 1537–1543 (1977).

34 Vahapassi, J. and Adlerkreutz, H.: Effect of continuous oral administration of estriol on corpus luteum function. Contraception *11:* 427–433 (1975).

35 Wang, D. Y. and Greenbaum, A. L.: The incorporation of (2-^{14}C) glycine and

(6-^{14}C) orotic acid into the acid soluble ribonucleotides and ribonucleic acids of the rat mammary gland. Biochem. J. *83:* 626–632 (1962).

36 WEBSTER, R. A. and HAMILTON, T. H.: Comparative effects of estradiol-17β and estriol on uterine RNA polymerase I, II and III *in vivo.* Biochem. biophys. Res. Commun. *69:* 737–743 (1976).

Prof. H. M. LEMON, Department of Internal Medicine, The University of Nebraska Medical Center, 42nd Street and Dewey Avenue, *Ohama, NB 68105* (USA)

Discussion

VAN KEEP: If the low incidence of breast cancer in women who had an early first pregnancy is, as you suggest, due to the influence of estriol on breast tissue, would it not be possible to prevent mammary carcinoma by giving girls of 17 or so a high dosage of estriol?

LEMON: I really do not know. Controlled clinical trials of estriol therapy would be needed to answer your question. VAHAPASSI and ADLERCREUTZ [Contraception *11:* 427–433 (1975)] have shown that 6 mg estriol raises the blood levels of free estriol to those seen in the first trimester of pregnancy. Any dosage above that only increases the amount of conjugated estriol, so a high dose would not be required. No toxic effects or amenorrhoea were reported in the pre-menopausal women treated. Estriol escapes from the uterine nuclear complex within a few hours, so the administration would have to be repetitive, but you then might see the same proliferative effect on the endometrium (or breast) as estradiol. There is some evidence that estriol, unlike estrone and estradiol, does not stimulate RNA polymerase 2 and 3 in the uterus, but whether or not this happens in the breast no one knows. Similarly, no one yet appears to know if a combination of two slightly different estrogen receptor complexes affects the nuclear chromatin like estradiol receptor complexes. Perhaps it is this mixture which shuts off the RNA polymerase activity and reduces tissue growth, preventing breast cancer in our treated rats.

THIJSSEN: RUDALI *et al.* [Eur. J. Cancer *11:* 39–41 (1975)] found that estriol pellets implanted under the skin of mice provoked mammary tumours in the same frequency after the same latency as estradiol pellets. They specifically point out that the estriol used did not contain impurities such as estradiol.

LEMON: The paper of RUDALI can be criticized. There are no histological data confirming that what was found were in fact cancers, and not lymphoid tumours. The dose administered, 400–600 μg/kg/24 h, was very high, and it should also be remembered that, using pellets, there was a continuous effect of estriol. The findings of RUDALI in this connection differ radically from those of other investigators, most of whom have reported great difficulty in inducing breast carcinoma in animals by estriol. It was for this very reason that, some 20 years ago, everyone lost interest in estriol as a possible carcinogen.

THIJSSEN: The $E_3/E_1 + E_2$ ratio in pre-menopausal patients in your study was obtained by the Brown method. I do not find the original Brown method very useful. As you know, several investigators, including ourselves, have not had the same results.

LEMON: About half of the publications which have appeared have, in fact, confirmed our findings. The differences may be explained by the way the population under study was stressed. One of the first things that changes in a woman under stress is her adrenal cortical output, and this is a precursor of estriol.

THIJSSEN: You mentioned that the estriol excretion is influenced by the production of DHEA and androstenedione; are you aware of the work of BROWN (Australia) who showed that the conversion of DHEA to estrogens – not only estriol – is very limited indeed: less than 0.005 %? [ADAMS, J. B. and BROWN, J. B.: Steroidologia *2:* 1–6 (1976)]. It is difficult to believe that at this rate this androgen has an important effect on the estriol levels.

LEMON: As you are aware, many other investigators have shown that adrenal-secreted Δ4-androstenedione is the major precursor of estrone biosynthesis by peripheral adipose and other tissues, and hence of estriol as well [SIITERI *et al.:* Proc. 4th Int. Congr. on Endocrinology, Washington, D. C., June 18–24, pp. 1237–1242, Excerpta Medica, Amsterdam 1972; RIZKALLAH, T. H.; TOVELL, H. M. M., and KELLY, W. G.: J. clin. Endocr. Metab. *40:* 1045–1056, 1975; MACDONALD, P. C., *et al.:* J. clin. Endocr. Metab. *27:* 1103–1111, 1967]. ACTH therapy in menstruating women may increase estrone and estradiol excretion rates without altering the estriol quotient [BARLOW, J. J.: J. clin. Endocr. Metab. *24:* 586–596 (1964)] while operative or other stress induces a two- to four-fold increase in the estriol quotient, especially in post-menopausal women [LEMON, J.: Am. med. Ass. *196:* 1128–1136 (1966); Cancer *23:* 781–790 (1969)]. In your own careful studies of estrogen metabolism in post-menopausal women with and without breast cancer [THIJSSEN: J. Steroid Biochem. *6:* 729–734 (1975)] in which no differences were observed either in blood production rates for androstenedione, for conversion of androstenedione to estrone, nor in estriol excretion between the two groups, one would like to know how recently mastectomy had been performed and what types of medication were administered to both groups of patients. Human and rabbit leucocytes do appear to have significant capabilities of conversion of dehydroepiandrosterone and Δ4-androstenedione to estriol and 16-epiestriol [LEMON, H. M. and REILLY, D.: Presented at Annu. Meet. of Endocrine Society, New York, June 18–20, 1975; PADRON, V. S.; STOHS, S. J., and LEMON, H. M.: unpublished data].

LAURITZEN: The ratio estriol/estrone + estradiol is really the metabolic reflection of the primary hormonal pattern. We hypothesized, many years ago, that the physiological role of estriol was to modify the effects of the primary hormones on the target organs, and that, because of this, for long-term therapy one should use impeded estrogens – such as estriol – or estrogens that can be metabolized into impeded estrogens [Fortschr. Med. *81:* 657 (1962)]. For the same reason I advocate estriol for patients who have had gynaecological cancer.

Is there any clinical evidence that estriol competes with estrone and estradiol at the level of the target organ on the breast tissue and the endometrium?

LEMON: There is not much evidence of this. One group in England, working with breast cancer patients, treated their patients first with estriol and then with tritium-labelled estradiol prior to biopsies or resection. They found that the take-up of estradiol was not blocked at all by the prior treatment with estriol. Unfortunately, the study was defective in that the group measured the gross uptake, much of which would, of course, have been non-specific and they failed to measure the nuclear concentration of the steroid-receptor complexes [DESHPANDE, N.; CARSON, P, and HORNER, J.: J. Steroid Biochem. *7:* 11–14 (1976)].

NEVES E CASTRO: We have treated cases of pre-menstrual mastodynia which did not react to other therapies, including progesterone, with 4–6 mg of estriol once a day. The results were very encouraging, and further studies are certainly merited.

LAURITZEN: Do high doses of estriol displace estrone, estradiol and androstenedione from receptors in subcutaneous fat tissue and in the fat tissue of the breast?

LEMON: Again I cannot give an adequate answer because too little is known. There is a metabolism of androgens to estrogens – all in low amounts – by human breast cancer [ADAMS, J. B. and WONG, M. S. F.: J. Endocr. *41:* 41–52 (1968)].

Recently very high concentrations of estriol conjugates were found in benign human breast cysts [RAJU, U.; GANGULY, M., and LEVITZ, M.: J. clin. Endocr. Metab. *45:* 429–434 (1977)]. These concentrations could only be explained by a local synthesis. Others have reported high concentrations of estrone and estradiol in human breast duct secretions [WYNDER, E. and HILL, P.: Lancet *ii:* 840 (1977)]. We have to consider that estrone and estradiol may be rapidly transformed into estriol by breast tissue as a means of relative inactivation, as a method of limiting the effect of estrogens. It is interesting that breast cancer tissue has a low rate of conversion.

NEVES E CASTRO: I was surprised to see that in rats you observed an anti-uterotropic effect when using estriol pellets. One would have expected the effect of continuous release from the pellets to have been similar to that seen when estriol is added to the drinking water, under which condition there is a marked stimulation of the uterus.

LEMON: I too found that difficult to explain. We were dealing with animals with a functioning ovary which is supplying estrone and estradiol. The estriol might have had an effect via the pituitary though this is difficult to understand since estriol is a poor inhibitor of pituitary function. It must be emphasized that estriol absorption was complete by 3 weeks, so that for 5 weeks there was no added steroid between implantations.

HASPELS: Has estriol ever been tried as a 'morning-after pill' to prevent a pregnancy after unprotected coitus?

LEMON: WOTIZ has tried this in a number of species, such as hamsters, rabbits and rats. The dosages were high: 200–1,000 μg/kg/day [WOTIZ, H. H. and SCUBLINSKY, A.: J. Reprod. Fert. *26:* 139–142; 143–150; 363–364; 365–368 (1971)]. He seems also to have done some work with women, but not much.

BAULIEU: Although I regard estriol as an interesting substance, and although I also think that your data are interesting, I disagree with your interpretation of them. Estriol is a weak estrogen, particularly in the human. It is, therefore, perhaps wrong to draw conclusions about its effects in animals such as rats who have little, if any, estriol.

You state that estriol has perhaps an anti-carcinogenic or anti-growth effect at the level of the estrogen receptor. I suggest it all depends on the experimental circumstances. Estriol may be anti-estradiol when competing for a single shot of the hormone, but its activity when permanently secreted or repeatedly injected is essentially estrogenic.

Although I am highly impressed by the epidemiological data showing correlations, I think there is still too little evidence to justify the use of estriol on a large scale as a prophylactic therapy for breast cancer.

LEMON: I feel that my hypothesis is not only defendable, but is being gradually confirmed. I appreciate your comments, though, I think we need sceptics like Professor BAULIEU as such criticism makes us sharpen our reasoning.

Front. Hormone Res., vol. 5, pp. 174–191 (Karger, Basel 1978)

Vascular Complications of Long-Term Oestrogen Therapy

D. McKay Hart, R. Lindsay and D. Purdie

Division of Obstetrics and Gynaecology, Stobhill Hospital, and Departments of Medicine and Gynaecology, Western Infirmary, Glasgow

Introduction

It is now well established that long-term oestrogen replacement has a protective effect on bone mineral content in both oophorectomised [Lindsay *et al.*, 1976] and post-menopausal women [Heaney and Recker, 1975]. Such an effect would justify long-term prophylactic oestrogen therapy for many women. Before such general treatment can be recommended however, it is important to assess the incidence and severity of any clinical and biochemical side effects which may result from long-term replacement with oestrogens.

Clinical Vascular Side Effects

In 1968 we commenced a long-term prospective trial of an oestrogen (mestranol) in patients who had had hysterectomy with or without bilateral oophorectomy, for non-malignant disorders [Aitken *et al.*, 1973]. Patients were randomly assigned to mestranol therapy ($2 \times 20 \mu g$/day) or a placebo, and the number of tablets taken by each individual patient calculated from postcards returned by the patients requesting further supplies of tablets.

During the first 1,000 woman-years of treatment with mestranol, the only significant clinical side effect was transient muscle cramps occurring in the legs (table I). These occurred almost 10 times more frequently in the group receiving active mestranol treatment than in the placebo-treated control group, and resembled the leg cramps commonly experienced in pregnancy. The total number of vascular incidents (cerebrovascular accidents,

Table I. Vascular incidents during mestranol and placebo therapy

	Mestranol	Placebo
Cramps in legs	19	2
Cerebral	3	2
Deep vein thrombosis	3	4
Myocardial infarction	1	3
Superficial vein thrombosis	6	7
Treatment years	1,030	950
Patients, n	207	202

myocardial infarctions, deep vein thromboses and superficial thrombophlebitis) was small and there was no detectable significant difference between placebo or mestranol groups (table I).

Clotting Mechanisms during Mestranol Therapy

146 patients taking mestranol and 121 taking placebo were screened for abnormal clotting function after 1–7 years of therapy.

Prothrombin time was measured by the method of QUICK *et al.* [1935] and was reported in seconds. Control assays were performed on pooled normal plasma. Partial thromboplastin time of plasma was estimated by the method of LANGDELL *et al.* [1953] using the Manchester Thromboplastin reagent.

Factor VII was measured by a modified single stage prothrombin time using a factor VII deficient human plasma (DADE) as substrate and expressing results as a percentage relative to cytosol. Factor X levels were determined by a modified stypven-time assay using factor X deficient plasma (DADE) and reported as a percentage relative to cytosol.

Only 1 patient in each group exhibited abnormal function using the routine screening tests of intrinsic (prothrombin time) and extrinsic (KCCT) pathways (table II). However, 24 patients taking mestranol had elevated factor VII compared with 7 in the placebo group. Comparing factor X results, 9 patients from the mestranol-treated group had elevated results compared with 3 in the placebo group.

Figure 3 shows the prothrombin times for both groups and figure 4 the partial thromboplastin times. When median and mean values were compared no significant difference could be detected between the 2 groups of

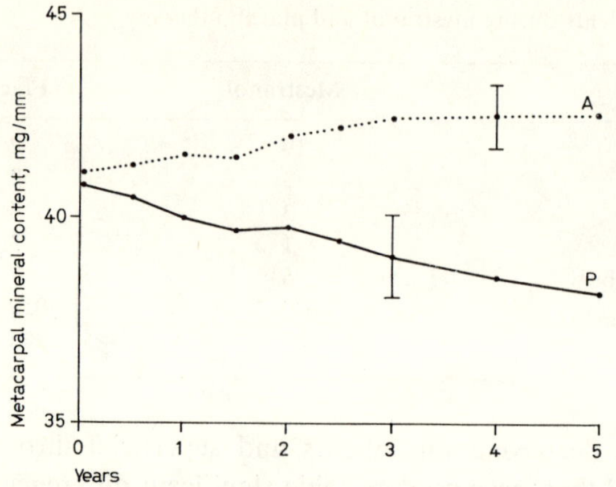

Fig. 1. Metacarpal bone mineral content during a 5-year double-blind trial of mestranol (A) versus a placebo (P). Bone mineral measured at mid-point of third right metacarpal by photon absorption technique (From Lancet, *ii:* 1039, 1976, by courtesy of the Editor).

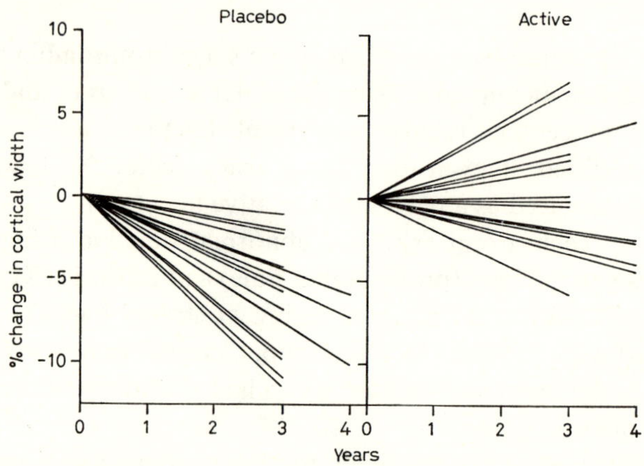

Fig. 2. Changes in metacarpal cortical width in individual patients taking mestranol or placebo. Lines are calculated from coefficients of individual measurements.

Table II. Abnormal screening tests during mestranol therapy

	Mestranol		Placebo	
	n	%	n	%
Abnormal 'screening' tests	1	0.7	1	0.8
Abnormal factor VII (> 125 %)	24	16.4	7	5.8
Abnormal factor X (> 125 %)	9	6.2	3	2.5
Patients, n	146		121	

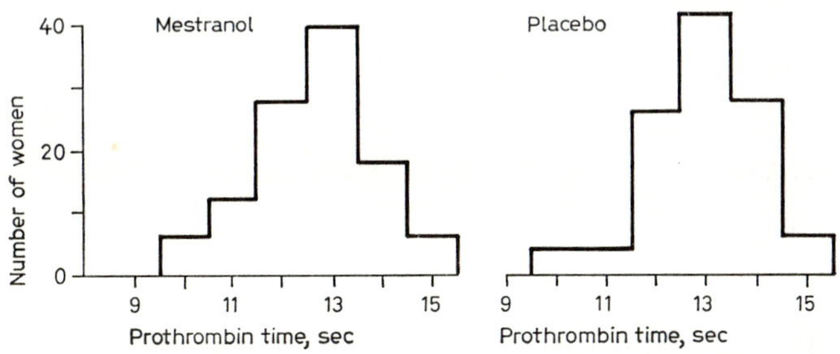

Fig. 3. Prothrombin times during mestranol and placebo therapy.

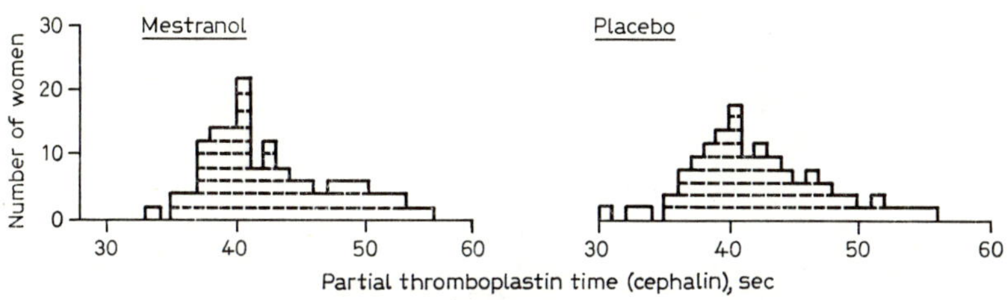

Fig. 4. Partial thromboplastin times during mestranol and placebo therapy.

patients. It is important to note that the scatter of results is similar in the two treatment groups.

When individual factors VII and X, both produced in the liver, were examined, mestranol produced a significant elevation of factor VII (fig. 5) but did not change factor X significantly (fig. 6). This confirms the view that oestrogens can influence the blood levels of components of blood coagulation, even when patients are taking a low daily dose. (The mean daily dose of mestranol actually taken by our patients was 25 μg.) Contrary

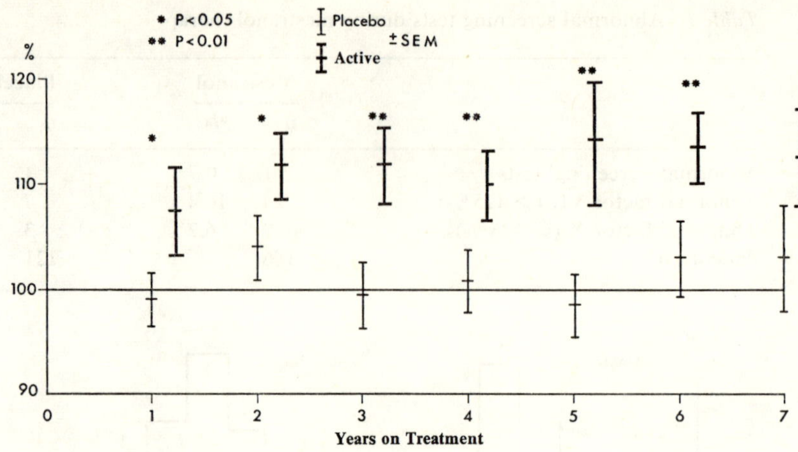

Fig. 5. Factor VII during mestranol and placebo therapy.

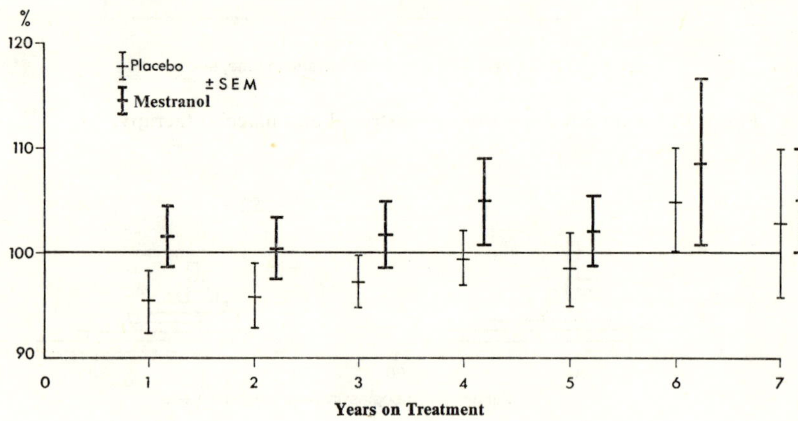

Fig. 6. Factor X during mestranol and placebo therapy.

to the findings of Ambrus [1976] none of our cases of thromboembolism had abnormal clotting function prior to the event. However, we would agree that changes in factors VII and X do not progress with time and random tests should indicate abnormalities. In other words, patients with normal values on one occasion tend to have repeatedly normal values while on therapy.

1 patient did have grossly abnormal screening tests on one occasion which inexplicably were normal when repeated 1 month later while still

Table III. Abnormalities of coagulation factors in 1 patient (B. M., 48 years) during mestranol therapy (duration: 5 years) with spontaneous reversion to normal, despite continuation of therapy

	Initial values		Repeat values	
	patient	control	patient	control
Prothrombin time, sec	8	12	13	14
KCCT, sec	24	38	34	41
Thrombin time, sec	7	6	3	4
Factor VII, %	>200	–	102	
Factor X, %	105	–	101	

on active treatment (table III). There was no evidence of a vascular incident at the time of the original sample and all attempts to elicit a cause for the abnormalities failed. The results have remained normal and the patient continues on therapy and in good health.

Clotting Indices on Oestriol

It has been suggested that the co-called 'natural' oestrogens and oestriol in particular have less effect on clotting mechanisms than the synthetic compounds such as mestranol [Davies *et al.*, 1975]. Oestriol has the advantage of not being converted *in vivo* to the more potent oestrogens oestrone and oestradiol. We have examined the effects on clotting function of incremental doses of oestriol hemisuccinate (Synapause, Organon) in a group of post-menopausal women.

27 women were given oestriol hemisuccinate 2 mg daily as continuous therapy. This daily dose was increased by 2 mg increments at 5-weekly intervals until the maximum 12 mg dose was obtained. Clotting activity was measured prior to starting therapy and immediately before each dosage increment. No abnormalities of prothrombin time or partial thromboplastin time were observed at any dose. However, at a daily dose of 6 mg oestriol and above, factor VII was significantly increased (fig. 7). Factor X was changed significantly only by a daily dose of 12 mg (fig. 8).

It is evident (table IV) that the number of abnormal tests tend to increase with the dose of oestriol prescribed, although in this particular study apart from the increased incidence of abnormalities at the 6 mg dose level (factor VII), only 8 and 12 mg resulted in a significant number of abnor-

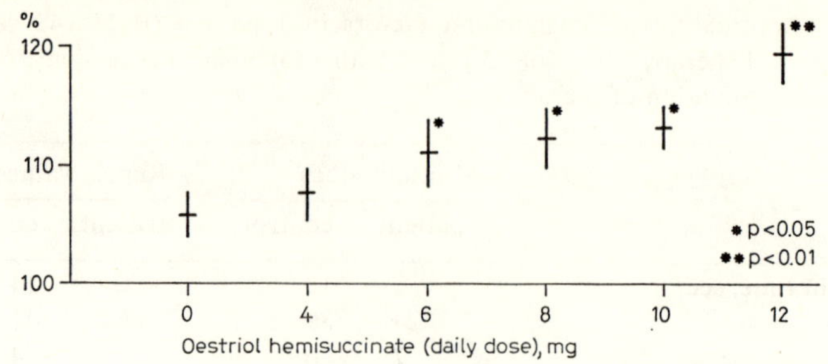

Fig. 7. Factor VII during incremental oestriol hemisuccinate therapy.

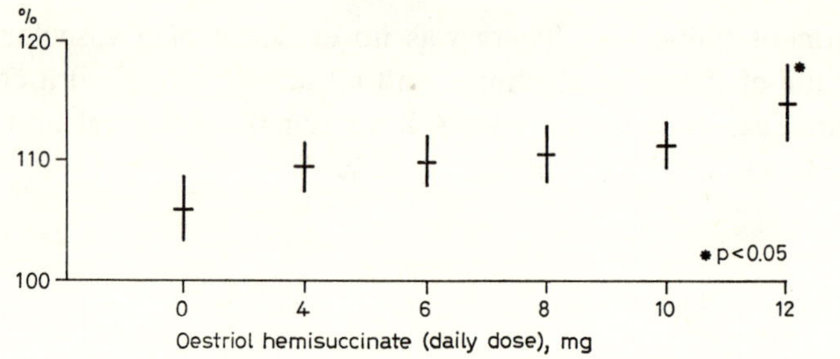

Fig. 8. Factor X during incremental oestriol hemisuccinate therapy.

Table IV. Number of abnormal clotting tests at different dose levels of oestriol hemisuccinate (patients: n = 27)

	Dosage, mg					
	0	4	6	8	10	12
Factor VII	1	1	8	6	4	12
Factor X	3	3	1	5	1	7

mal results. Curiously at 10 mg/day fewer abnormalities were recorded than at the 2 immediately lower doses.

These findings suggest that the effects of either a 17-alkylated oestrogen (mestranol) and a so-called natural oestrogen (oestriol hemisuccinate) are essentially similar and that any differences are merely dose dependent. Such differences will be of considerable importance if a lower equivalent dose of the natural oestrogen (in terms of its effect on clotting function) has an equal effect on symptoms and bone loss to the relatively higher dose of the synthetic compound.

Mestranol Therapy and Serum Lipids

We have described previously some preliminary results indicating that mestranol therapy lowered serum cholesterol but increased serum triglycerides [AITKEN *et al.*, 1971]. Long-term observation of our group of patients confirms the reduction in cholesterol and the increase in triglycerides resulting from mestranol treatment (fig. 9). These changes are highly significant (fig. 10) and are similar to the observations of ROBINSON and LEBEAU [1965].

The rise in triglycerides is consistent with the effect of oestrogens on very low density lipoproteins (VLDL) and it may be that the overall fall in cholesterol is indicative of a negative effect on high density lipoproteins (HDL). The effect of mestranol on cholesterol levels is dose dependent and is biphasic. Low doses of mestranol (< 20 μg/day) result in an orerall increase in serum cholesterol and only a higher daily intake of mestranol

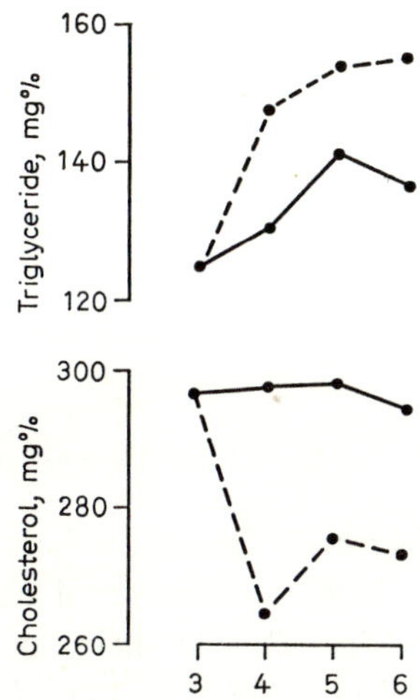

9

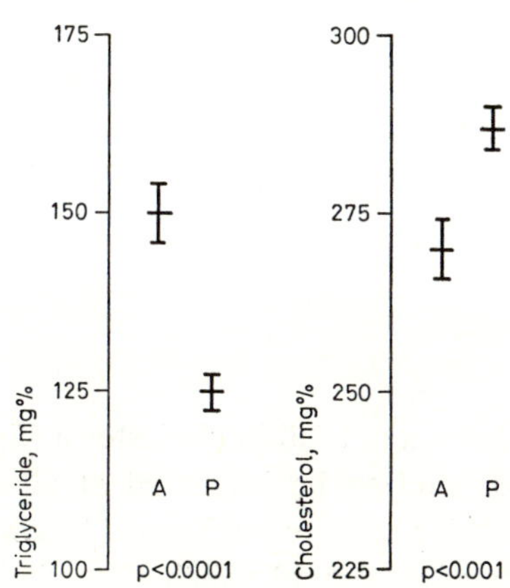

10

Fig. 9. Changes in fasting cholesterol and triglyceride levels during mestranol and placebo therapy.

Fig. 10. Significance of changes in fasting lipid levels.

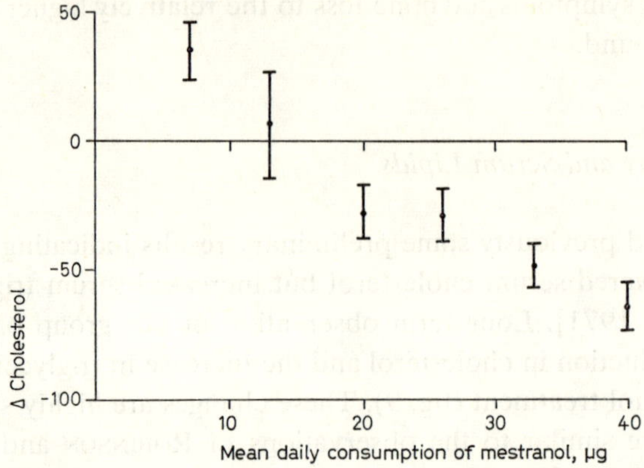

Fig. 11. Relation between mean daily dose of mestranol and fasting serum cholesterol.

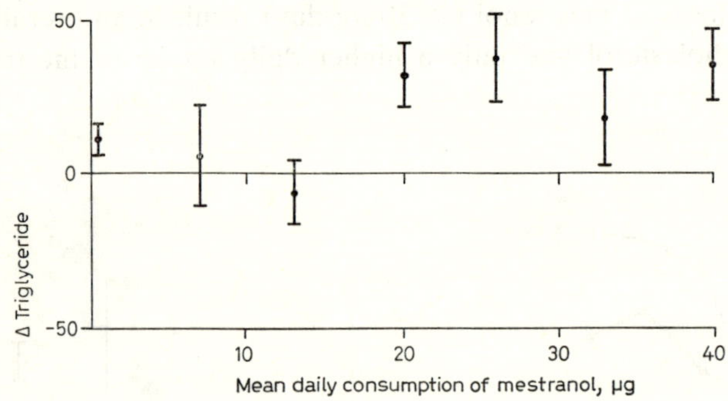

Fig. 12. Relation between mean daily dose of mestranol and fasting serum triglyceride.

produces a net fall in cholesterol (fig. 11). Triglyceride levels, on the other hand, tend to be increased at all daily dose levels of mestranol above 15 μg/day and are unchanged below this value (fig. 12). It is likely that the doses of mestranol which raise cholesterol (and possibly HDL) and leave triglycerides unchanged and therefore probably have beneficial effects on fat metabolism, are equally effective in reducing bone mineral loss [Aitken *et al.*, 1974] (fig. 13).

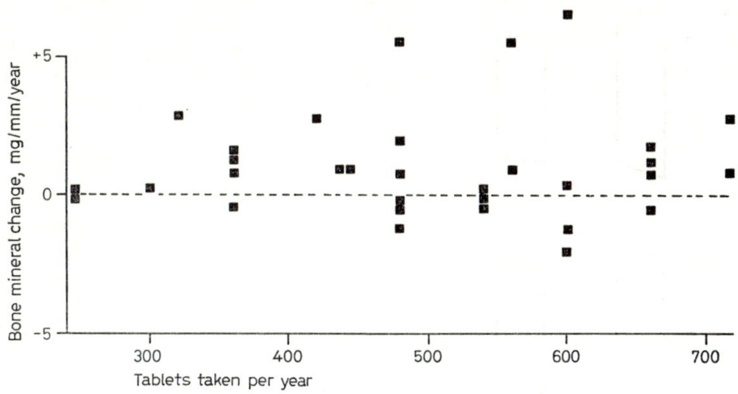

Fig. 13. The relationship between mestranol dosage, estimated as number of 20 μg tablets taken per year, and the changes in metacarpal mineral content.

Blood Pressure Changes during Mestranol Treatment

Throughout our mestranol study, arterial blood pressure was measured annually after 15 min resting in the supine position. Figure 14 compares mean diastolic and systolic blood pressures in 120 patients who had completed 3 years treatment with mestranol or a placebo.

In slim patients there was little change in either diastolic or systolic blood pressure during the 3 years of follow-up. Mestranol therapy did not significantly effect blood pressure in this group. However, in obese patients (body weight > 25 % above the average weight for age and height at the start of the trial), there was a significant increase in both diastolic and systolic blood pressure in the group taking placebo tablets. There was *no* significant change in blood pressure in patients taking oestrogen, confirming the short-term studies of von Eiff [1975].

Weight Changes

Since, in our patients, blood pressure changes were more closely allied to obesity than to oestrogen treatment, we examined the effect of oophorectomy and subsequent oestrogen replacement on weight itself. An initial cross-sectional study of 450 patients was carried out. Patients were reviewed 3 months, 3 years and 6 years after hysterectomy with or with-

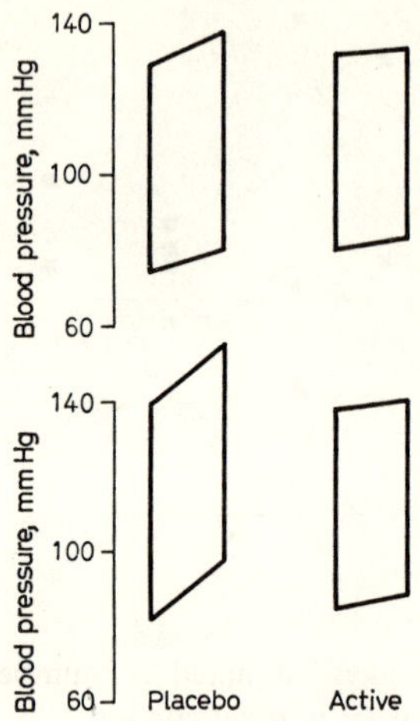

Fig. 14. Systolic and diastolic blood pressures at the start and end of 3 years of treatment with mestranol or a placebo. Groups have been divided into obese patients and those of normal body weight or less.

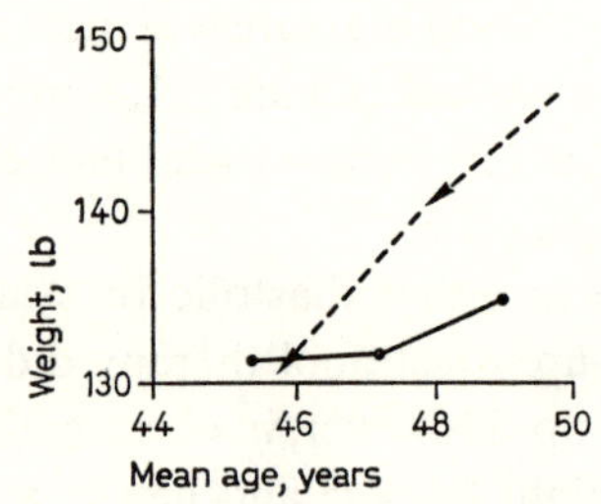

15

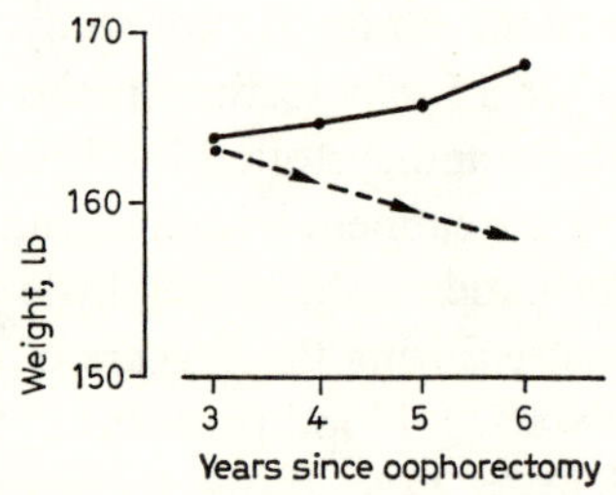

16

Fig. 15. Comparison of mean weights in three groups of patients of similar ages at operation reviewed 2 months, 3 years and 6 years after hysterectomy (–) or hysterectomy plus bilateral oophorectomy (– – –) carried out pre-menopausally.

Fig. 16. Weight changes in mestranol (▲) and placebo (●) treated obese patients followed for 3 years.

out bilateral salpingo-oophorectomy, carried out pre-menopausally. Body weight increased with age in oophorectomised patients (fig. 15) but remained steady in those patients whose ovaries had been conserved until about age 47 years when mean weight increased; this being roughly the mean age at which ovarian function would tend to be diminished.

Mestranol therapy, instituted 3 years after oophorectomy surprisingly reversed the trend toward increased weight in obese patients, while those

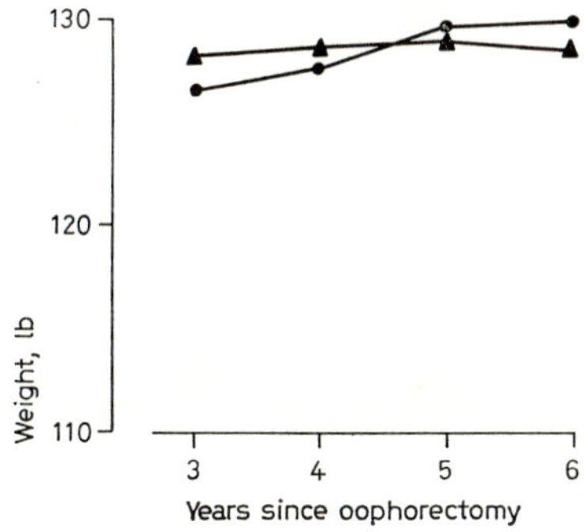

17

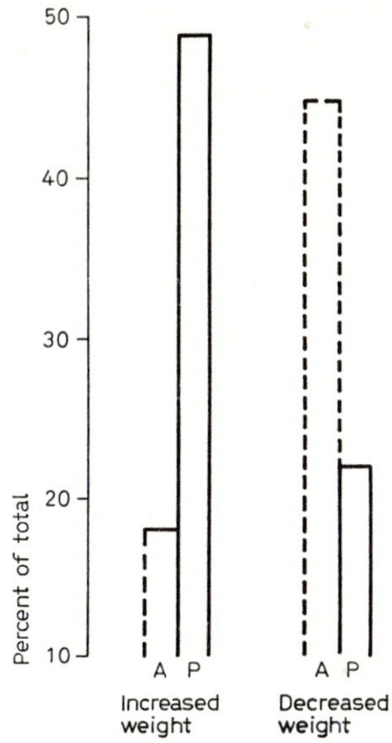

18

Fig. 17. Weight changes in mestranol (▲) and placebo (●) treated patients of normal weight followed for 3 years.

Fig. 18. Percentage of total patients who changed weight during mestranol (A) or placebo (P).

patients taking placebo continued to put on weight (fig. 16). In patients of normal weight there was no subsequent weight change when oestrogen or placebo treatment was begun. However, when the total number of patients in the trial were surveyed, the majority of those who lost a significant amount of weight were taking mestranol, while the majority who gained weight were taking the placebo preparation (fig. 18).

Measurement of mid-triceps skinfold thickness was carried out at 6-monthly intervals during the first 3 years duration of treatment. Measurements were performed in duplicate by the same observer. Despite the high error involved, a significant increase in skinfold thickness at this site occurred in the placebo-treated group, confirming the trend toward weight increase observed in this group. Consistent with this, and the weight loss already referred to in mestranol-treated patients, there was a slight fall in skinfold thickness, and a significant decrease was observed when the actively treated group were compared with those taking placebo tablets.

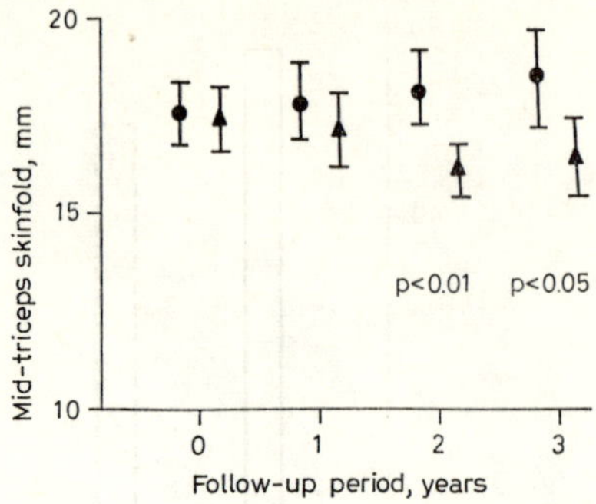

Fig. 19. Changes in triceps skinfold thickness during 3 years treatment with mestranol or placebo therapy. ● = Placebo; ▲ = active.

Table V. Factors considered in calculating myocardial infarction risk score

1	Family history positive for ischaemic heart disease
2	Cigarette smoking
3	Diastolic BP > 95 mm Hg
4	Obesity (> 5 kg from expected weight)
5	Fasting blood sugar > 90 mg%
6	Fasting cholesterol > 270 mg%
7	Fasting triglycerides > 150 mg%

Risk of Ischaemic Heart Disease

Collecting such data in a group of patients taking oestrogen therapy, could be used to assess the relative risks of developing ischaemic heart disease [Sznajderman and Oliver, 1963]. Risk scores were calculated for each patient using an established method (table V) and the average scores for each group of patients calculated. It is clear from the data in figure 20 that the myocardial infarction risk score following hysterectomy carried out pre-menopausally was lower than in patients of similar age reviewed 1 year more after hysterectomy and bilateral oophorectomy. Long-term mestranol therapy did tend to reduce the mean risk score of the latter group (fig. 21) but this was only significant within the first 5 years of operation. The area from which these patient populations come is within the Glasgow district, where a notoriously high prevalence of ischaemic heart disease, particularly in men, but also obvious when the female population is com-

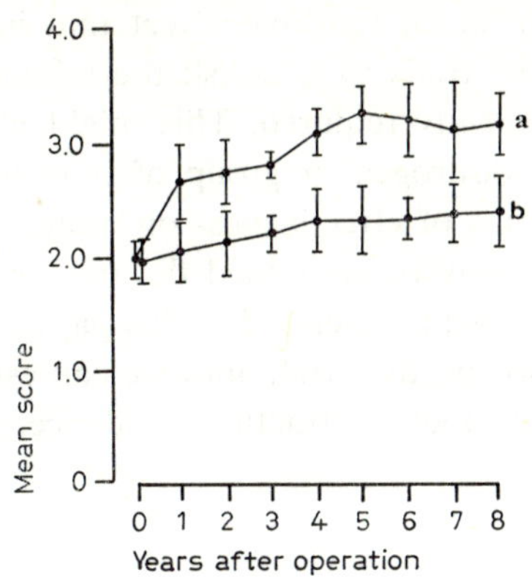

20

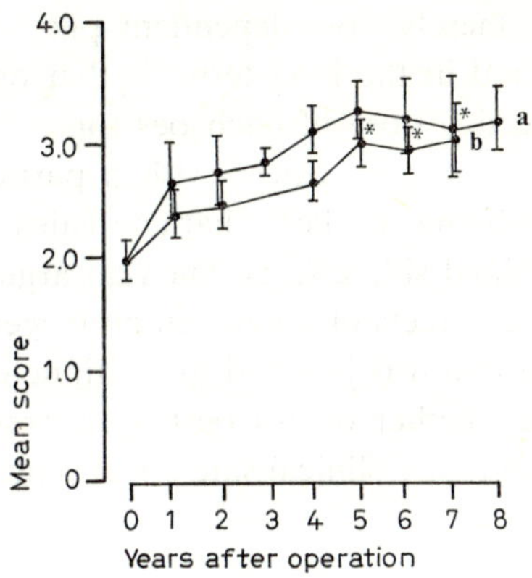

21

Fig. 20. Coronary risk score following hysterectomy or hysterectomy plus bilateral salpingo-oophorectomy. a = Following hysterectomy and oophorectomy; b = following hysterectomy.

Fig. 21. Effect of mestranol therapy on coronary risk score in oophorectomy. a = Following hysterectomy and oophorectomy; b = mestranol therapy following oophorectomy. * Means not significantly different.

pared to that of other areas. Despite this high risk, we have recorded insufficient data about the incidence of myocardial infarction within our group of patients to allow us to draw definite conclusions about the effect of oestrogens (table I).

Discussion

If long-term oestrogen therapy is to be provided for those women who are at risk from post-menopausal osteoporosis, then some prospective assessment of the possible side effects of such therapy must be made. This we have attempted to do using our prospective trial of mestranol against a placebo preparation. The results on bone loss indicate a continuing preventive effect of oestrogen over at least the first 5 years of therapy (fig. 1). In this population no serious clinical side effects have as yet become obvious. However, persistent elevation of factor VII, a vitamin K dependent clotting factor of hepatic origin, has been observed. We have also shown that the effects of oestriol succinate on factors VII and X are dose dependent and would suggest that previously reported differences between the synthetic oestrogens such as mestranol and the so-called natural oestrogens are merely dose-dependent phenomena. Since the final effect that is required in the long term, is that on bone, trials to establish the minimum effective dose of each oestrogen are urgently required. This might allow rational replacement with a particular oestrogen or group of oestrogens exhibiting the best characteristics in terms of effectiveness on bone, with minimal side effects. Such an argument may be continued throughout the other effects of long-term oestrogen treatment, described in this paper. We feel that it is justified to continue our prospective trial, until we can establish whether or not oestrogen prevention reduces fracture incidence, and what if any clinical side effects emerge.

Summary

Long-term low dose oestrogen therapy has a protective effect on bone mineral content in the post-menopausal or castrated female. As yet the only obvious clinical side effect of such therapy has been transient leg muscle cramps.

Several biochemical side effects could be observed. Low dose mestranol caused a persistent elevation of factor VII and a dose-dependent increase in both factors VII and X was observed using oestriol hemisuccinate. Such effects are more likely to be dose-related than related to the type of oestrogen prescribed. Effects of oestrogens on lipids, and cholesterol in particular, may be dose-related also.

Changes in blood pressure in post-menopausal women are more likely to be related to obesity than to oestrogen treatment which would seem to have a protective effect against weight increase. No marked changes in the mean risk score for ischaemic heart disease could be detected during oestrogen treatment.

References

Aitken, J. M.; Lorimer, A. R.; Hart, D. McK.; Lawrie, T. D. V., and Smith, D. A.: The effects of oophorectomy and long-term mestranol therapy on the serum lipids of middle-aged women. Clin. Sci. *41:* 597 (1971).

Aitken, J. M.; Hart, D. M., and Lindsay, R.: Oestrogen replacement therapy for prevention of osteoporosis after oophorectomy. Br. med. J. *iii:* 515–518 (1973).

Aitken, J. M.; Hart, D. M.; Lindsay, R., and MacDonald, E. B.: Oestrogen dosage in the treatment of post-menopausal osteoporosis. Medikon Int. *3:* 3–5 (1974).

Ambrus, J. L.: Oestrogens and clotting factors, chap. 13; in van Keep, Greenblatt and Albeaux-Fernet Consensus on menopause research (MTP Press, Lancaster 1976).

Davies, T.; McNicol, G. P.; Fieldhouse, G.; Gallagher, J. C., and Nordin, B. E. C.: A comparison of the effects of oestriol succinate and ethinyl oestradiol on blood coagulation, platelet function and fibrinolysis in post-menopausal women;

in VAN KEEP and LAURITZEN Estrogens in the post-menopause. Front. Hormone Res., vol. 3, pp. 185–198 (Karger, Basel 1975).

EIFF, A. W. VON: Blood pressure and estrogens; in VAN KEEP and LAURITZEN Estrogens in the post-menopause. Front. Hormone Res., vol. 3, pp. 177–184 (Karger, Basel 1975).

HEANEY, R. P. and RECKER, R. R.: Oestrogen effects on bone remodelling at menopause. Clin. Res. *23:* 535 (1975).

LANGDELL, R. D.; WAGNER, R. H.; BRINKHOUSE, K. M., and HILL, C. N. C.: Effect of anti-haemophilic factor on one-stage clotting tests. J. Lab. clin. Med. *41:* 637 (1953).

LINDSAY, R.; HART, D. M.; AITKEN, J. M.; MACDONALD, E. B.; ANDERSON, J. B., and CLARKE, A.: Long-term prevention of post-menopausal osteoporosis by oestrogen. Lancet *i:* 1038–1041 (1976).

QUICK, A.; STANLEY-BROWN, M., and BANCROFT, F. W.: A study of the coagulation defect in haemophilia and in jaundice. Am. J. med. Sci. *190:* 501 (1935).

ROBINSON, R. W. and LEBEAU, R. J.: Effect of conjugated equine oestrogens on serum lipids and the clotting mechanism. J. Atheroscler. Res. *5:* 120 (1965).

SZNAJDERMAN, M. and OLIVER, M. F.: Spontaneous premature menopause, ischaemic heart disease and serum lipids. Lancet *i:* 962 (1963).

Dr. D. MCKAY HART, Division of Obstetrics and Gynaecology, Stobhill General Hospital, *Glasgow G21 3UW* (United Kingdom)

Discussion

RAKOFF: According to your findings, perhaps mestranol does not deserve its bad reputation. On what basis did you decide on the dosage of 40 μg?

MCKAY HART: In 1968, when this trial began, this dosage was chosen somewhat arbitrarily. It was similar to the dose of mestranol in oral contraceptives, and so we felt that it would be safe as well as effective. As far as the bad reputation of mestranol is concerned, it is interesting to note that whilst over the years a number of patients on placebo therapy have stopped taking their tablets because of side effects of various kinds, we have not had a single one who asked to come off mestranol. At this dosage there appears to be a satisfactory control of vasomotor symptoms. Only very, very occasionally has there been a patient complaining of dyspareunia.

KOPERA: To what extent was this a blind trial? Half the women were on a placebo for a very long time. Did your patients know about this possibility in advance?

MCKAY HART: All the women involved in this trial were regular visitors to the clinic, some for as long as 9 years, which undoubtedly explains the extremely low drop-out rate. I saw all the patients at the outset, and it was explained to them that we did not know for sure that their bone density would benefit from the therapy. The study was in fact a randomized one, the trial numbers being all allocated by a colleague who did not know which patients were receiving which numbers.

UTIAN: On the question of weight, we have treated 50 oophorectomized women for over 2 years and had seen no significant change of weight at all.

McKAY HART: This may be explained by the fact that you did not divide the patients into 'obese' and 'slim' at the outset in quite the same way that we did. I am not sure that if we had simply checked the weight of all patients after a period of 2 years we would have found a very marked difference for the group as a whole. It is only in the obese patients that a clear change was seen.

UTIAN: Was the fall you found in cholesterol values statistically significant? I think this is an important point, because whilst there is accumulating evidence that the ovary plays a protective role as far as coronary risk is concerned, most studies indicate that it is not possible for the same protection to be provided after oophorectomy by the administration of exogenous estrogens. We have recently conducted a large study and the data analysed so far show absolutely no significant reduction in cholesterol levels.

McKAY HART: The fall in the group as a whole over the 3-year period was indeed significant. We are now planning to analyse the cholesterol and triglycerides data according to the patients being obese or slim and see what this does to the significance.

LAURITZEN: I am a little bit surprised, Dr. UTIAN, by your finding that exogenous estrogen does not decrease plasma cholesterol. This seems to be at variance with the results of many other investigations.

UTIAN: At the same time there are publications which confirm my findings. I would mention, for instance, a recent publication of ROSENBERG, ARMSTRONG and JICK in which the authors report that when looking at the incident rates of myocardial infarction, they could find absolutely no evidence that estrogens protect against coronary thrombosis [New. Engl. J. Med. *294:* 1256 (1976)].

McKAY HART: My point was not to show that there would be an improvement of coronary score with estrogen therapy, but that there was no worsening of the score, and I think that this as such is very important.

NORDIN: Your risk score is a simple summation of factors and not so much a weighting, is that correct?

McKAY HART: Yes, that is right.

LAURITZEN: Did you see any significant increase in the risk of venous thrombosis? GOW and MACGILLIVRAY [Br. med. J. *ii:* 73–77 (1971)] found a high incidence, 16%, in post-menopausal women treated with mestranol using as little as 20 μg/day for 4 months only.

McKAY HART: This was indeed a very strange finding. These authors treated only 25 patients and, if mestranol really had such an effect, we in our much larger study, should have had many such episodes.

UTIAN: I discussed this paper with Professor MACGILLIVRAY, and he told me that his was a very selected population in that risk patients had not been excluded from therapy. The fact that at least 3 of the 25 patients in the study had previously suffered deep venous thrombosis is quite a significant point.

LAURITZEN: I think it is important that this be stated here, because this paper is often quoted to show that there is a specific risk of thromboembolism for 'estrogen' treated patients in the post-menopause, a risk very much greater than the small one incurred with the risk of oral contraceptives. This is obviously not so.

VOIGT: The average pre-treatment cholesterol values of your patients were rather high, very close to 300 mg, while the triglyceride levels were in the lower range of normal. Could the drop you observed in cholesterol levels under therapy be explained by a change in the women's eating habits?

MCKAY HART: We have not analysed people's diet patterns as time for this was lacking. However, knowing the characteristics of the population from which this particular group of women was drawn, I doubt very much that there has been any change in this respect. It is not a population which easily changes its habits.

I wonder if there is any significant, perhaps even fundamental, difference between the effect on cholesterol and triglyceride levels of the combined estrogen and progesterone preparations, such as are used for the oral contraceptive and for replacement therapy around the time of the menopause, and therapy with estrogens and intermittent progestagen used to produce a periodic shedding of the endometrium. As far as I know, no one has published data on this.

NEVES E CASTRO: I wonder whether any studies on clotting factors will bring us closer to knowing which estrogens, in what dosage, have an effect on thromboembolism. Would it not be much more wise to look at viscosity, blood flow, and perhaps the vascular wall?

MCKAY HART: I think you have a very valid point there, but at the same time you have to agree that these things are incredibly difficult to study.

KLOPPER: How do the quantities of estriol succinate that you used compare with estriol as such?

MCKAY HART: 2 mg of estriol succinate is equivalent to about 1.3 mg of estriol.

NORDIN: For the record, I must say that I disagree with you about the relative errors of densitometry and morphometry. The error of densitometry in our hands was about 2–4 %, whereas the multiple metacarpal morphometry has a relative error of about 1 %.

MCKAY HART: We used morphometry of a single metacarpal and cannot comment on the errors in multiple measurement.

LEMON: How were the oophorectomized patients randomized?

MCKAY HART: The oophorectomy was done before the patients came to our clinic. Initially we saw all patients from two hospitals who had had an hysterectomy; but after a number of years the load became too great, and we then saw only the oophorectomized patients. In 1968, when we started this mestranol trial, we were seeing all patients who would come. Whether a patient had had a hysterectomy or a hysterectomy plus oophorectomy was at the discretion of any of about ten different clinicians. The neat age matching between the hysterectomized patients and those who were hysterectomized *and* oophorectomized occurred, in fact, by chance.

LEMON: Did you see any cases of breast cancer among these women who were followed for such a long time?

MCKAY HART: No, surprisingly we did not, but I really think that a much larger study would be needed before we could say that this is the result of therapy.

Front. Hormone Res., vol. 5, pp. 192–202 (Karger, Basel 1978)

The Effect of Different Regimens of Oestrogens on the Clotting and Fibrinolytic System of the Post-Menopausal Woman

M. Thom, M. Dubiel, V. V. Kakkar and J. W. W. Studd

King's College Hospital, Denmark Hill, London

Introduction

Oestrogen-progestogen preparations were used for contraceptive purposes for 15 years before epidemiological data on their thrombo-embolic complications were revealed [Medical Research Council, 1967; Vessey and Doll, 1969]. The changes in platelet function, fibrinolysis and blood coagulation are believed to be due to oestrogen rather than the progestogen component of oral contraceptives [Poller, 1973]. The increased risk of death incurred by the use of oral contraceptives is estimated to be 43 in 49,681 woman years [Vessey *et al.*, 1977], occurring in women who are heavy smokers. Oestrogen therapy in the post-menopausal women may have similar risks as her coagulation factors, V, VII, X, XI and XII [Hamilton *et al.*, 1974] and fibrinogen levels [Ogston and Ogston, 1966] are already increased with age.

Bucher *et al.* [1973], Shormorken [1970] and Sano *et al.* [1971] have demonstrated an age-dependent increase in platelet adhesion and aggregation, perhaps because of an increased sensitivity of the platelets to ADP [Sano *et al.*, 1971]. Age-specific factors such as the decreased blood flow in response to exercise [Redisch *et al.*, 1962] and the increased liability of the elderly to intimal vascular damage may predispose them to the development of thrombo-embolic disease.

Hormone replacement therapy must therefore be shown to be safe in the older woman, so it may be used for its beneficial and preventative effects on the same scale as the contraceptive pill. This study was implemented to compare the effects of various hormone regimens on coagulation, fibrinolysis and platelet function.

Table I. Patients

Treatment group	Patients, n	Mean age, years	Mean years post-menopausal	Mean parity
I Conjugated equine oestrogens	6	49.8 (47–54)	1.5 (1–2)	1.8 (0–4)
II Oestradiol valerate	6	51.5 (47–56)	2.2 (1–4)	1.3 (0–3)
III Piperazine oestrone sulphate	5	51.0 (48–54)	1.4 (1–3)	1.4 (0–3)
IV Sequestrial ethinyl oestradiol with megestrol acetate	5	52.4 (43–60)	4.0 (1–8)	2.0 (0.4)
V Ethynodiol diacetate	6	51.3 (47–55)	3.7 (1–7)	1.5 (1–2)
VI Oestradiol and testosterone implant	5	43 (35–48)	3.6 (1–8)	2.4 (1–4)

Ranges are indicated in parentheses.

Patients and Methods

33 symptomatic, post-menopausal patients were recruited for the study from the Dulwich Hospital Menopause Clinic. Each patient had a history taken and a general and pelvic examination with cervical cytology.

No patient had a history of thrombo-embolic disease and all were within 20% of the average weight for their age and height. None had contraindications to hormone replacement therapy so they were randomly allocated to 1 of 6 treatment groups. None of the patients had received any form of hormone therapy. Table I shows the characteristics of the patients.

Group I. Conjugated equine oestrogens (Premarin) 1.25 mg orally daily for 3 weeks out of four.

Group II. Oestradiol valerate (Progynova) 2 mg orally daily for 3 weeks out of four.

Group III. Piperazine oestrone sulphate (Harmogen) 3 mg orally daily for 3 weeks out of four.

Group IV. Serial 28, sequestrial ethinyl oestradiol 10 μg with megestrol acetate 1 mg added for 5 days in each cycle.

Group V. Ethynodiol diacetate, 0.5 mg daily, orally.

Group VI. Oestradiol 50 mg and testosterone 100 mg by subcutaneous implant with norethisterone 5 mg daily for 5 days of each cycle to produce a regular withdrawal bleed in each of the 2 patients with a uterus.

Blood Samples

Venous blood samples were taken in a plastic syringe before and after 3 months of therapy between 9 and 11 a. m. and put into citrated bottles for coagulation, fibrinolytic and platelet aggregation studies. The blood was spun at 1,000 rpm for 10 min to produce platelet-rich plasma, and at 3,000 rpm for 10 min to produce platelet-poor plasma for coagulation studies. The patients were not fasting when the samples were taken.

Platelet counts were performed on a Coulter Thrombocounter.

Platelet aggregation was measured using platelet-rich plasma with a Born Mark IV aggregometer, in ADP (Sigma Chemicals Co. Ltd., Surrey, England) and adrenalin in 0.5 and 1.0 μM concentrations and in thrombin 0.2 U/l (Parke-Davis Co., Pontypool, Wales).

Plasminogen concentration was measured by an esterolytic method described by SCULLY *et al.* [1977].

Fibrin degradation products were measured in serum by the method of MESKEY *et al.* [1966] using commercial reagents (Wellcome, England).

Keolin-cephalin clotting time was estimated by the method of HARDISTY and INGRAM [1966] using cephalin kindly supplied by Lean Poller, National Reference Laboratory for Anticoagulant Control Reagents, Manchester.

Anti 10-A was measured using activated 10-A Sets and plasma CEF reagent (Sigma) by the method of YIN *et al.* [1973].

Results

Coagulation

There was no significant change in the levels of anti-Xa after treatment (fig. 1).

Ethynodiol diacetate produced a significantly reduced ($p < 0.05$) kaolin-cephalin clotting time after 3 months of treatment and were the only group to demonstrate any change (fig. 2).

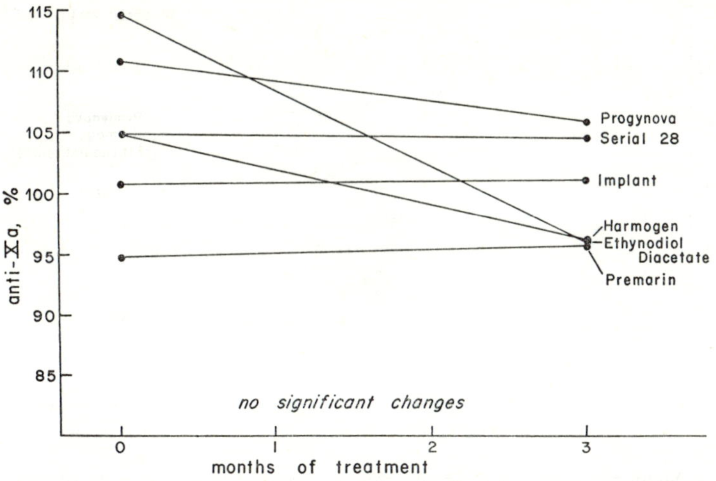

Fig. 1. Mean anti-Xa before and after treatment.

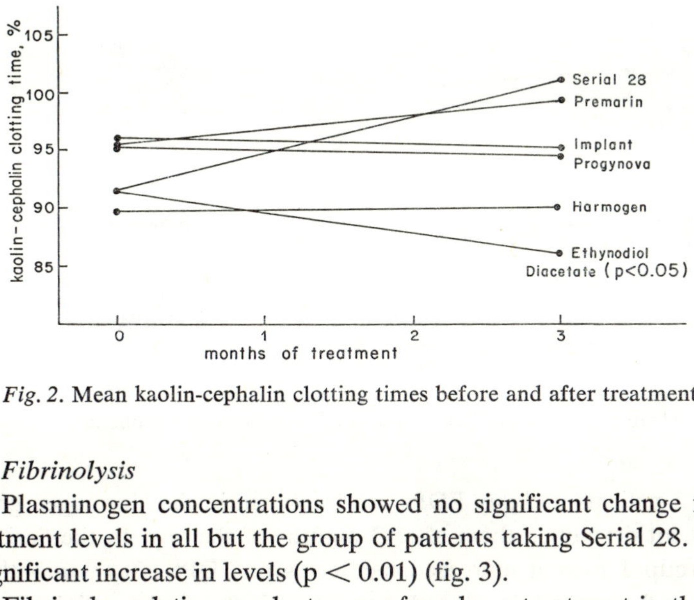

Fig. 2. Mean kaolin-cephalin clotting times before and after treatment.

Fibrinolysis

Plasminogen concentrations showed no significant change from pre-treatment levels in all but the group of patients taking Serial 28. They had a significant increase in levels ($p < 0.01$) (fig. 3).

Fibrin degradation products were found pre-treatment in three of the Premarin-treated groups. After 3 months they disappeared in 2 patients, were reduced in the third and appeared in one other patient. In the Progynova group of patients they were found pre-treatment in 3 patients, increased after 3 months in two and decreased in the third.

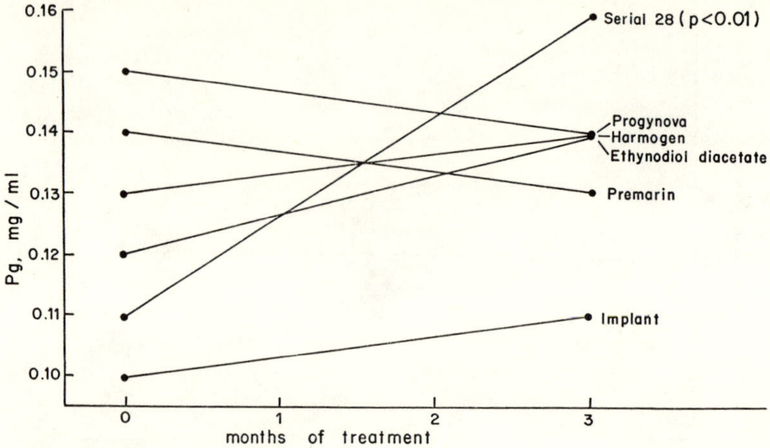

Fig. 3. Mean plasminogen concentration before and after 3 months of treatment.

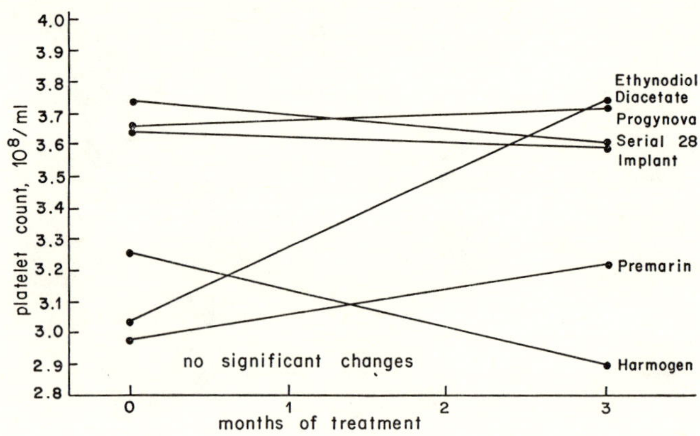

Fig. 4. Mean platelet count before and after 3 months of treatment.

Only 2 patients had any FDPs pre-treatment in the Harmogen group, and 1 patient developed them after 3 months of therapy. In the Serial 28 treated group 1 patient increased her levels of FDPs after 3 months of treatment and a second developed them. 2 patients on ethynodiol diacetate developed FDPs and a third increased her level. 2 patients who were given implants had FDPs before treatment; 1 had none after 3 months and the level in the other patient remained unchanged. No patients in this group developed FDPs during therapy.

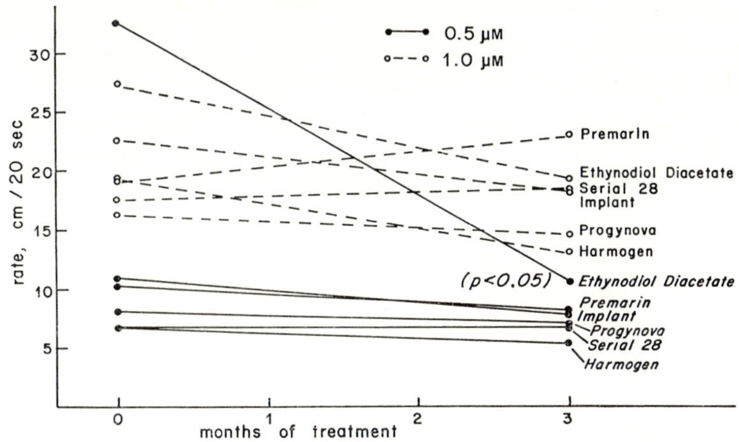

Fig. 5. Platelet aggregation in AOP; change in rate after 3 months' therapy.

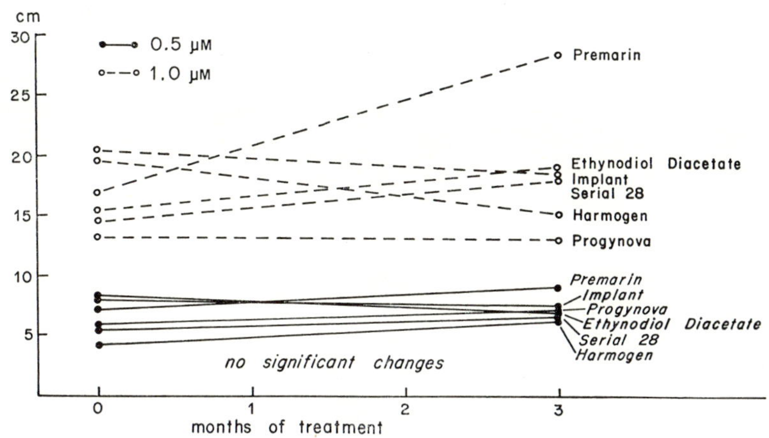

Fig. 6. Platelet aggregation in AOP; change in extent after 3 months' therapy.

Platelet Function

No group showed any statistically significant changes in platelet count after 3 months of therapy (fig. 4). Platelet aggregation studies revealed no significant changes in patients having treatment with either Progynova or an implant (fig. 5, 6).

After 3 months of Harmogen therapy patients had a significant decrease in the rate of platelet aggregation measured in ADP of 1.0 μM solution ($p < 0.05$) and also an increase in the extent of aggregation in adrenalin of 1.0 μM solution ($p < 0.05$). Patients treated with Serial 28 for 3 months developed an increase in the extent of their platelet aggregation in adrenalin of 0.5 μM solution ($p < 0.05$). Patients on ethynodiol diacetate therapy showed a decrease in the extent of platelet aggregation in adrenalin in both 0.5 and 1.0 μM solutions ($p < 0.05$). No group showed a significant change in platelet aggregation measured in thrombin.

Discussion

GAW and MACGILLIVRAY [1971] showed that although mestranol successfully treated symptoms of oestrogen deficiency, it was associated with a 16% incidence of venous thrombosis. VON KAULLA *et al.* [1975] and COOPE *et al.* [1975] showed that conjugated oestrogens were associated with the hypercoagulable state finding an increase after treatment in factors VII and X and a decrease in the prothrombin time. However, NOTELOVITZ and GRIEG [1975] found no change in prothrombin time. Neither COOPE *et al.* [1975] nor NOTELOVITZ and GRIEG [1975] found any change in the kaolin partial thromboplastin time. The only group who showed a change in the kaolin-cephalin clotting time in our study after 3 months of therapy were those on continuous ethynodiol diacetate ($p < 0.05$).

Age does not affect the platelet count [HAMILTON *et al.*, 1974], but BELLER *et al.* [1972] using mestranol, and NOTELOVITZ and GRIEG [1975] using Premarin both showed a fall in the platelet count. No group showed any significant alteration in platelet count in this study.

BELLER *et al.* [1972] found plasminogen concentrations decreased after therapy with 80 μg of mestranol. BONNER [1975] also showed that the plasminogen concentration fell in response to treatment with Syntex-Menophase. Patients on Serial 28 in our study were the only ones who showed a significant change in plasminogen levels after therapy and they showed an increase ($p < 0.01$). VON KAULLA *et al.* [1975] found that there was a shift towards lower anti-thrombin III activity in women on conjugated oestrogens. No group showed any significant alteration in anti Xa levels after therapy in our study.

Fibrin degradation products were found to increase after 6–9 months of 22 μg/day of ethinyl oestradiol by DAVIES *et al.* [1976]. In our study a total of 9 patients developed FDPs after therapy so demonstrating that there had been an increase in fibrinolytic activity.

COOPE *et al.* [1975] and NOTELOVITZ and GRIEG [1975] found platelet aggregation unaffected by conjugated oestrogens, but Davies *et al.* [1976] found it decreased in ADP, and increased in adrenalin after ethinyl oestradiol therapy and increased in collagen after oestriol succinate, which also increased platelet aggregation in adrenalin. In this study no significant changes in platelet aggregation were found in any of the groups of patients on Premarin, Progynova or implants.

The extent of platelet aggregation in adrenalin was significantly increased in patients after therapy with Harmogen in 1.0 μM solutions, and Serial 28 in 0.5 μM solution. Patients on ethynodiol diacetate therapy showed a decrease in the extent of platelet aggregation in adrenalin of 0.5 and 1.0 μM solutions.

After 3 months of Harmogen therapy patients showed a significant decrease in the rate of platelet aggregation in ADP of 1.0 μM solution.

The incidence of mortality in women on the pill is related to age, therefore it might be thought that hormone replacement therapy in the post-menopausal woman would be even more dangerous. However, the oestrogens generally used are not the synthetic mestranol or ethinyl oestradiol used in the birth control pill and the dose required as assessed by oestrogenic potency is approximately 25% of that required to inhibit ovulation in the younger woman. There are also favourable changes which occur in blood lipids with a decrease in cholesterol and β-lipoprotein levels shown by Magnani and Moore [1976], although there is frequently an associated elevation in plasma triglycerides.

Whatever may be the significance of isolated changes in laboratory investigations within the whole complex coagulation cascade and their interrelation with vessel compliance with increasing age and blood lipid changes, it is becoming clear from epidemiological studies that the incidence of thrombotic events is not increased in women taking the so-called natural oestrogens for the climacteric syndrome.

This study does not reveal any change towards hypercoagulability in any of the parameters with the treatment regimens studied, but it must be emphasised that the numbers involved are small. This supports the epidemiological data of the Boston Collaborative Study [1973] which failed to show any increase in the incidence of thrombotic episodes in post-menopausal women, taking oestrogen therapy. However, it is obvious that more patients need to be studied and currently factor anti-Xa assays are being performed by an autoanalytic technique on all patients in our menopause clinic, as factor anti-Xa has been shown to be predictive in recognising patients at risk of thrombosis on the oral contraceptive pill [Sagar *et al.*, 1976].

Summary

The effects of six different regimens of hormone replacement therapy on coagulation, fibrinolysis and platelet aggregation were studied in 33 symptomatic, postmenopausal women. Studies were performed before and after 3 months of therapy in the six treatment groups, Premarin, Progynova, Harmogen, Serial 28, ethynodiol diacetate and a testosterone and oestradiol implant.

No changes in any parameter were detected following administration of Premarin, Progynova or an implant of testosterone and oestradiol. Serial 28 produced a significant increase in plasminogen concentration and increased the extent of platelet aggregation in 0.5 μM adrenalin solution. Harmogen produced an increase in the extent of platelet aggregation in 1.0 μM adrenalin solution, but a decrease in the rate of platelet aggregation in 1.0 μM solution. Ethynodiol diacetate caused a significant decrease in the kaolin-cephalin clotting time but decreased the extent of platelet aggregation in 0.5 and 1.0 μM solutions of adrenalin.

FDPs developed following therapy in 9 of the 33 patients. These were evenly distributed between the groups. No patients had any change in anti-Xa concentration, platelet count or platelet aggregation in thrombin.

References

BELLER, F. K.; NACHTIGALL, L., and ROSENBERG, M.: Obstet. Gynec. *39:* 775 (1972).

BONNER, J.: in CAMPBELL The management of the menopause and post-menopausal years, p. 321 (MTP, London 1975).

Boston Collaborative Drug Surveillance Program: Lancet *i:* 1399 (1973).

BUCHER, U.; ROBERT, Y., and RIEDWYL, H.: Thromb. diath. haemorrh. *29:* 671 (1973).

COOPE, J.; THOMPSON, J. M., and POLLER, L.: Br. med. J. *iv:* 139 (1975).

DAVIES, T.; FIELDHOUSE, G., and MCNICHOL, G.: Thrombosis and Haemostasis *35:* 403 (1976).

GAW, S. and MACGILLIVRAY, I.: Br. med. J. *ii:* 73 (1971).

HAMILTON, P. J.; ALLARDYCE, M.; OGSTON, D., and DOUGLAS, A. S.: J. clin. Path. *27:* 980 (1974).

HARDISTY, R. M. and INGRAM, G. I. C.: Bleeding disorders (Blackwell, Oxford 1966).

MAGNANI, H. N. and MOORE, B.: Post-grad. med. J. *52:* suppl. 6, p. 55 (1976).

Medical Research Council: Br. med. J. *2:* 355 (1967).

MESKEY, C.; KEINER, G. J., and JOHNSON, A. J.: Blood *28:* 1 (1966).

NOTELOVITZ, M. and GRIEG, H. B. W.: S. Afr. med. J. *49:* 101 (1975).

OGSTON, C. M. and OGSTON, D.: J. clin. Path. *19:* 352 (1966).

POLLER, L.: in POLLER Recent advances in blood coagulation, p. 181 (Churchill, London 1973).

REDISCH, W.; MECKELER, K., and STEELE, J. M.: in BLUMENTHAL Medical and clinical aspects of ageing, pp. 453–459 (Columbia University Press, New York 1962).

Royal College of General Practitioners: Oral Contraception Study, (Pitman, London 1974).

SAGAR, S.; STAMATAKIS, J. D.; THOMAS, D. P., *et al.:* Lancet *i:* 509 (1976).
SANO, T.; BOXER, M. G. J.; BOXER, L. A., and YOKOYAMA, M.: Thromb. Diath. haemorrh. *25:* 524 (1971).
SCULLY, M. F.; LANE, D. A.; SAGAR, S.; THOMAS, D. P., and KAKKAR, V. V.: Thrombosis Haemostasis *37:* 162 (1977).
SHORMORKEN, H.: Acta med. scand. *188:* 339 (1970).
VON KAULLA, E.; PROEGMUELLER, W., and VON KAULLA, K. N.: Am. J. Obstet. Gynec. *122:* 688 (1975).
VESSEY, M. P.; MCPHERSON, K., and JOHNSON, B.: Lancet *ii:* 731 (1977).
VESSEY, M. P. and DOLL, R.: Br. med. J. *ii:* 651 (1969).
YIN, E. T.; WESSLER, S., and BUTLER, J.: J. Lab. clin. Med. *81:* 298 (1973).

Dr. J. W. W. STUDD, Department of Obstetrics and Gynaecology, King's College Hospital, Denmark Hill, *London SE5 9RS* (United Kingdom)

Discussion

UTIAN: Your comparison of the effects of various estrogens on clotting is very interesting. The only type of preparation not included in your study seems to be an unconjugated artificial estrogen. Your findings are in agreement with those of NOTELOVITZ and GREIG [J. Reprod. Med. *16:* 87–90 (1976)], who were the first to suggest a difference between the action of natural estrogens and that of artificial ones, such as ethinyl estradiol and, probably, mestranol.

STUDD: Our findings are at variance with those of POLLER and COOPE [The management of the menopause and post-menopausal years, pp. 313–320 (MTP, Lancaster 1976)]. One point to consider is that most of the studies which have been done into the side effects of the so-called synthetic estrogens have involved preparations in which the estrogen content is very much higher than is considered necessary for the treatment of menopausal problems. The dose in the oral contraceptive Serial 28, for instance, was some ten times greater. I do not know of any published work on the clotting factors of 10 μg of ethinyl estradiol, which I personally would consider about the right dose for menopausal therapy.

NORDIN: I understood that the relationship between the various clotting parameters and clinical episodes was still to be established. Have you seen any patients on estrogen therapy with thrombo-embolic phenomena?

STUDD: Not really. There were 3 patients with superficial thrombosis in our first series of patients; these were receiving Menophase. The data from the menopause clinic have not all been evaluated yet, but there has been no deep venous thrombosis and no pulmonary embolus.

NORDIN: I had only 2 true cases of deep venous thrombosis in many hundreds of patient-years, and no pulmonary embolism. There were several cases of superficial thrombophlebitis, but I am now beginning to ignore this.

LAURITZEN: I too have not seen any serious complications of this type in more than 2,000 patient-years. BONNAR saw a decrease in factor Xa using a sequential type combination of norethisterone and mestranol [The management of the menopause and post-menopausal years, pp. 321–329 (MTP, Lancaster 1976)]. Did you not see any effect of factor Xa of progestogens?

STUDD: No, we did not. BONNAR's findings with that preparation were more disturbing than the findings of others.

Front. Hormone Res., vol. 5, pp. 203–219 (Karger, Basel 1978)

Estrogens, Estrogen Receptors, and Biological Responses in Experimental Animals

T. E. Shellenberger and D. M. Sheehan

Department of Health, Education and Welfare, Food and Drug Administration, National Center for Toxicological Research, Jefferson, Ark.

Introduction

The human population is exposed to estrogenic compounds at various dose levels and exposure periods. Small amounts of naturally occurring estrogens in animal or plant products, or residues of anabolic agents, such as DES, are ingested involuntarily by large segments of the population for long periods of time. In this case, particularly sensitive processes, such as fetal development, may be involved. Voluntary exposures occur via use of oral contraceptives, hormone replacement therapy, and treatment of appropriate mammary or prostate tumors. In these instances, doses are higher, but the duration of exposure may vary from a few days to years.

The variability in mode, extent, and duration of exposure to various estrogens poses many regulatory questions regarding human risk. An example is the problem of determining whether estrogens are carcinogenic due to their estrogenic properties or whether another mechanism, such as metabolic activation, must be invoked. Likewise, the possible existence of a threshold for carcinogenicity requires thorough investigation. Thus, there is an explicit need for improvement in existing or for development of new protocols, animal models, etc. to determine estrogenicity and carcinogenicity as well as to determine the impact of exogenous estrogens on endogenous estrogen levels and hormones under estrogen control.

The Hormone (Estrogen) Program at the National Center for Toxicological Research has been designed to investigate these questions in a series of integrated projects including studies of certain biochemical parameters, *in vivo* bioassays for carcinogenic and estrogenic activity in rodents, and metabolism studies in a variety of systems including nonhuman primates

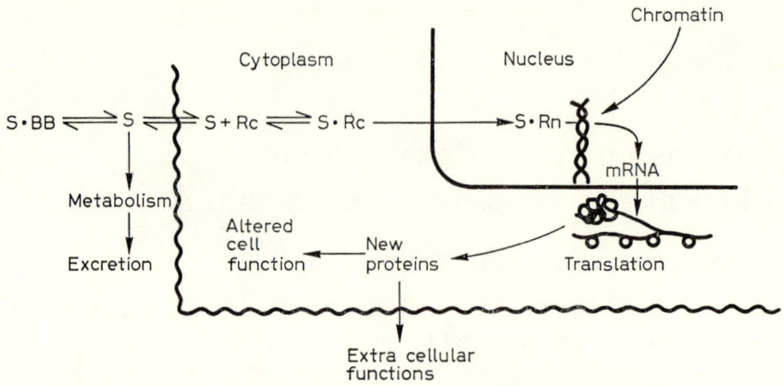

Fig. 1. Mechanism of estrogen action. This model shows the major steps of estrogen action where: S = any estrogen; S · BB = S bound to blood protein; S · Rc = S bound reversibly to a cytoplasmic receptor; and S · Rn⁻ is the S · Rc translocated to the cell nucleus.

and humans. These projects can be conceptually related at one level to the mechanism of estrogen action which is depicted in figure 1. Estrogens, designated S, are transported in the blood compartment where they bind reversibly to blood proteins (S · BB), undergo metabolism and excretion either as parent chemical or as conjugated metabolites, or penetrate the cells of the target tissue. In the target cell, the estrogen is first bound reversibly to a cytoplasmic receptor – forming an estrogen receptor complex (S · Rc) – which, following activation, is translocated to the cell nucleus ($S \cdot RN^-$) where it can bind to chromatin resulting in gene activation. The messenger RNA (mRNA) transcribed from these genes is transported to the cytoplasm and translated into new proteins which are then responsible for altered cell function.

Within the context of this model, results from several current projects will be discussed sequentially: (a) interaction at the extracellular level of E_2 and diethylstilbestrol (DES) with plasma proteins; (b) preliminary data on estrogen effects on RNA, and (c) estrogen-mediated uterine weight gain, reflective of altered cell function, and (d) tumor incidence and pathologic findings related to estrogen-induced reproductive tract lesions.

Protein-Hormone Equilibrium Binding Studies

Protein binding was measured using methods described by PEARLMAN and CREPY [13]. Briefly, [^{3}H] estradiol, or [^{3}H] diethylstilbestrol (Ameri-

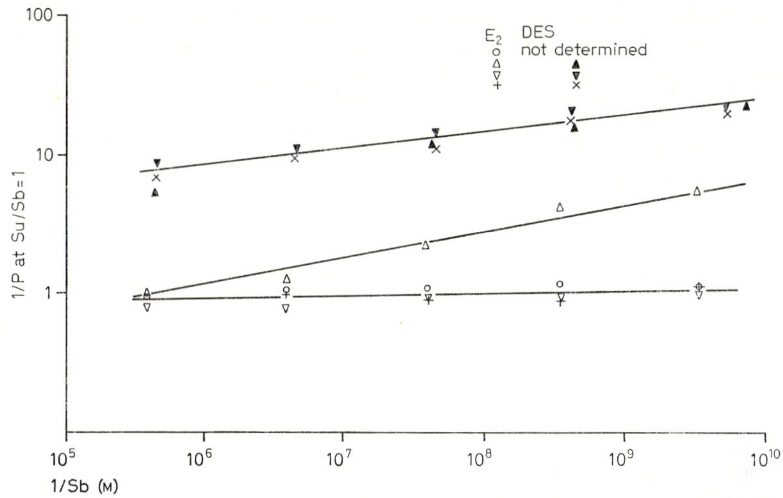

Fig. 2. Binding affinity of E_2 and DES for plasma and BSA. Various plasma or BSA samples were titrated against several fixed concentrations of ^3H E_2 or ^3H DES. Values for Su/Sb were determined with the Sephadex assay and the affinity index, 1/P at Su/Sb = 1 is plotted against 1/Sb.

can Searle, 40–60 Ci/mmol) was added to vials (1.5 × 4.5 cm) containing 200 mg Sephadex G-25 (coarse) preswollen in 1 ml of 0.155 M sodium phosphate, pH 7.0. Approximately 100,000 cpm were added per vial. Other components included plasma or protein and nonradioactive carrier to a final volume of 2.0 ml. After gentle shaking for 30 min at room temperature, the vials were removed and the Sephadex was allowed to settle for 30 min. Samples of the supernatant were taken for determination of ^3H. In this assay, hormone is partitioned between the external and internal phases of the Sephadex. Protein, however, is excluded from the internal volume. The amount of hormone found above protein-free controls is a measure of binding. Details of the theory and data treatment may be found in PEARLMAN and CREPY's work [13], while further details of the binding studies will be reported elsewhere [17, 21].

The results from these studies indicate that the two estrogens, E_2 and DES, display distinctly different binding patterns to plasma proteins (fig. 2). The binding affinity ($^1/p$ at Su/Sb = 1, where Su and Sb are, respectively, unbound and bound steroid [13]) was measured over a wide range of input hormone concentrations. Several important features emerge from these data (fig. 2). First, DES binds 10- to 20-fold better than E_2 to all

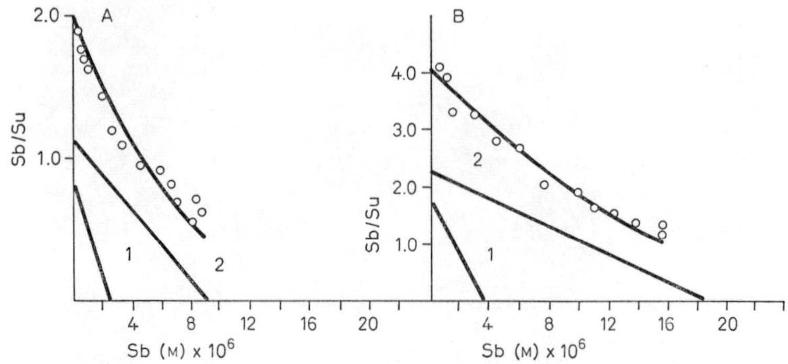

Fig. 3. Scatchard plots of DES-BSA binding. The interaction of DES with BSA was determined at 66.7 μg/ml BSA (panel A) or 133 μg/ml BSA (panel B). The points are the experimental data while the lines through the points were obtained by assuming two binding components, indicated by the lines marked 1 and 2. The constants obtained from these lines are: Panel A, line 1, $K_A = 3.3 \times 10^5$ M^{-1}, n = 2.5; line 2, $K_A = 1.2 \times 10^5$ M^{-1}, n = 12; and Panel B, line 1, $K_A = 4.4 \times 10^5$ M^{-1}, n = 2; line 2, $K_A = 1.3 \times 10^5$, n = 9.

plasmas from nonpregnant animals. Second, the binding affinity of DES changes as a function of the hormone concentration, indicative of site heterogeneity. Third, the binding of DES to plasma and to purified bovine serum albumin (BSA) appears indistinguishable. This is consistent with data which suggests that albumin is the primary blood binder of several steroids [4]. Fourth, although estradiol binds much more effectively to plasmas containing the high affinity binding protein α-fetoprotein (α-FP) than to nonpregnancy plasmas, DES demonstrates no such apparent discrimination.

All of these points have been confirmed by further experiments. Figure 3 is a Scatchard plot which demonstrates characteristics of the DES-BSA interaction. The site heterogeneity seen in figure 2 is manifested as a curved line in the Scatchard plot of DES-BSA binding (fig. 3). This line can be resolved into two components with $K_A \sim 4 \times 10^5$ M^{-1}, n = 2 and $K_A = 1.3 \times 10^5$ M^{-1}, n = 9–12. Thus, DES shows low affinity, high capacity binding to BSA, which is also the type of binding of steroids with BSA. Because of the low solubility of E_2, a complete Scatchard analysis for E_2-BSA binding cannot be obtained. However, SANDBERG *et al.* [15] report that E_2-BSA binding has an $nK = 1 \times 10^5$ M^{-1} which can be compared to the sum of the nK's ($\sim 20 \times 10^5$ M^{-1}) calculated from figure 3. These calculations are consistent with the 10- to 20-fold difference be-

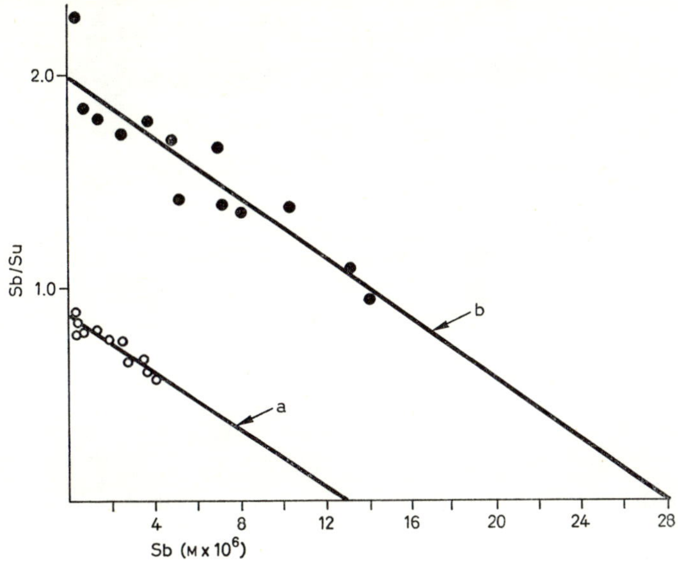

Fig. 4. Binding of DES to plasma. Scatchard plot of DES binding to plasma obtained from 85-day-old female rats. The protein concentration was 166 μg/ml (line a) and 333 μg/ml (line b). The values for K_A and n (in moles DES/mg protein) are: Line a, $K_A = 6.9 \times 10^5\ M^{-1}$, $n = 7.7 \times 10^{-8}$, line b, $K_A = 7.0 \times 10^5\ M^{-1}$, $n = 8.5 \times 10^{-8}$.

tween E_2 and DES binding shown in figure 2. Figure 4 demonstrates in a Scatchard plot the binding of DES to a plasma sample. Although the site heterogeneity seen with BSA is not apparent, this may be due to the presence of other components in plasma or, more likely, to the greater scatter in the data. The $K_A = 7 \times 10^5\ M^{-1}$ is somewhat higher than that seen with DES-BSA, but the data points may represent predominantly the higher affinity component ($K_A \sim 4 \times 10^5\ M^{-1}$) seen in figure 3. Assuming that plasma is 50% serum albumin [17] combined with the value of $\sim 8 \times 10^{-8}$ M DES/mg protein for plasma (fig. 4) it can be calculated that n = 10–11 for rat serum albumin in plasma. This value is quite close to n = 9–12 found for purified BSA. Therefore these data show that the binding of DES in a simple system of BSA and buffer may be useful to obtain information for modeling the delivery of DES to target tissues.

The fourth point demonstrated in figure 2, that DES does not bind with high affinity to plasma containing α-FP, has been investigated in more detail. Rather than pregnancy plasma, rat amniotic fluid was chosen as the

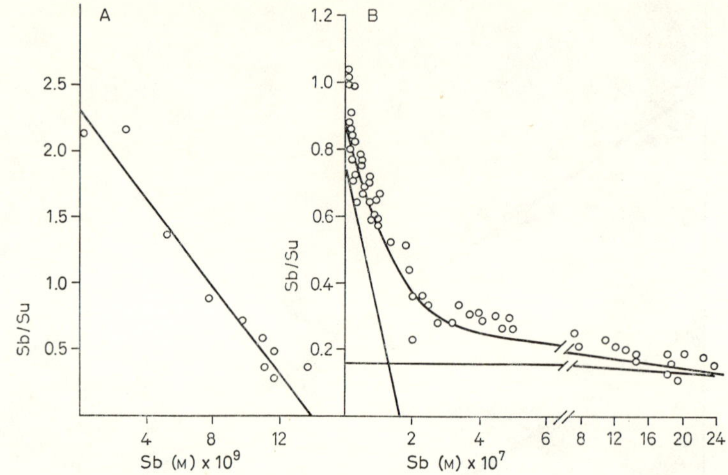

Fig. 5. Binding of E_2 and DES to α-fetoprotein. Amniotic fluid containing α-FP was titrated with E_2 (panel A) or DES (panel B). The calculated α-FP concentrations were 1.4×10^{-8} M (panel A) and 5.2×10^{-8} M (panel B). With E_2, the $K_A = 1.6 \times 10^8$ M^{-1} and n = 1.0. With DES, the more complex binding curve has been resolved into two components. The points represent data, while the line through them has been calculated from the two component model. The high affinity component has $K_A = 0.5 \times 10^7$ M^{-1} and n = 3.

source of α-FP at reasonable concentrations (1 mg/ml). Figure 5 shows a Scatchard analysis of E_2 and DES binding to rat α-FP. The binding of E_2 is used in this laboratory to provide a control valve for comparison with DES binding. In agreement with other investigators [1, 20], we find that the $K_A = 1 \times 10^8$ M^{-1} and n = 1 mol E_2/mol α-FP. These same preparations, however, show more complex binding curves with DES. There are at least two components present. When these data are fitted with a two-component model (lines), the higher affinity component ($K_A = 0.5 \times 10^7$ M^{-1}) is found to have three binding sites for DES. A three-component model (not shown) will also fit the data adequately, and again shows three binding sites for DES. The low affinity component is probably serum albumin, which is found in appreciable quantities in amniotic fluid. This supposition is based on two observations. First, the low affinity component shares many binding characteristics in common with BSA. Second, addition of increasing amounts of BSA to amniotic fluid caused progressive changes in the low affinity, but not the high affinity binding component. Experiments are in progress to purify α-FP from amniotic fluid for further binding studies. These will hopefully allow a resolution of the nature of the low affinity component as well as a more precise definition of the characteristics of DES binding to α-FP. There has been considerable disagreement in the literature concerning the interaction of DES with α-FP. Several authors [9, 18, 19] indicated that DES failed to bind at all, while others [8, 10, 14] detected some affinity of DES for α-FP. Most of these investigators used competition methods to examine binding. These are prone to several types of errors since they are optimized for detection of E_2 binding and not DES binding. The characteristics of DES-α-FP binding reported here have been obtained in a systematic manner by directly determining the binding under equilibrium conditions. SANU *et al.* [16], also using equilibrium conditions, have seen extensive binding of DES but did not determine values for K_A or n. In addition to carrying out binding experiments with purified α-FP, competition experiments of several types are underway to verify information on the number and types of sites for estrogens on α-FP.

With these data for DES binding to serum albumin and α-FP, attempts will be made to derive a predictive model for the delivery of estrogens to target tissues. With such a model, for instance, it might be possible to quantitatively examine the effect differences in binding between DES and E_2 have on the availability of these hormones to sensitive tissues of the rat fetus and pup. These studies may provide information related to the *in utero* carcinogenic properties of estrogens.

RNA Studies

Our group is also attempting to develop a complementary DNA (CDNA) probe to assess normal and aberrant hormone action. As part of this project, we have been collecting data about the effects of estrogens on nucleic acid in the rat uterus. Ovariectomized Sprague-Dawley rats (60–90 days old) were implanted (s. c.) with estradiol in Silastic tubing or paraffin pellets. At various times, animals were sacrificed, uteri were removed and stored in liquid N_2 and plasma was prepared for RIA determinations of estradiol. RNA was extracted with an alkaline phenol-chloroform procedure, purified through repeated extractions, precipitated with alcohol, and stored at –20 °C. Aliquots were passed over oligo-dT cellulose columns for analysis of poly-A containing RNA. Details of the above procedures will be reported elsewhere [2, 3].

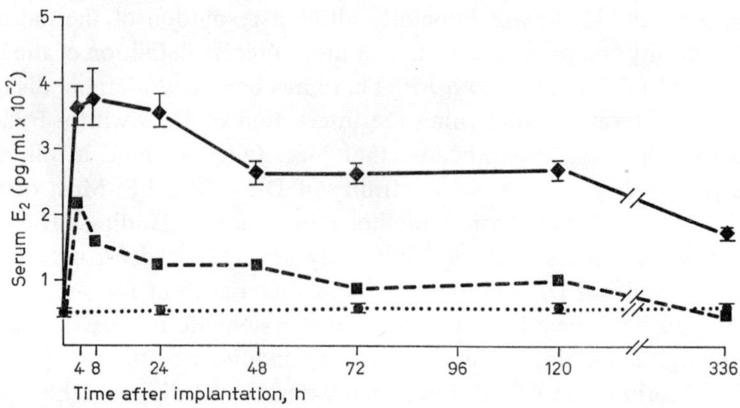

Fig. 6. Serum estradiol levels in implanted rats. Ovariectomized rats were sacrificed at intervals after implantation in either Silastic capsules (♦; 250 μg E_2) or paraffin pellets (■; 120 μg E_2). Control animals (●) received no E_2. Serum E_2 was analyzed by double antibody RIA. Vertical lines represent SEM.

Most investigators using the rat uterine system have been concerned with the short-term effects of estrogens; and have, therefore, administered the hormone by injection. We chose to characterize techniques for continuous delivery of constant amounts of hormone for several reasons. First in addition to the short-term responses, our interests also include the effects of long-term exposure. Second, the rapid rise and fall in estrogen levels resulting from injection might seriously limit the type, duration and interpretation of a variety of experiments. Third, continuous delivery of hormone offers more flexibility in the superimposition of other hormone treatments. Our first set of experiments, then, were concerned with the delivery of estradiol.

Figure 6 shows plasma estradiol levels achieved with paraffin or Silastic implants. Both show an initial pulse of E_2, but plasma estradiol levels in the paraffin implanted animals fall subsequently while Silastic implants continue to release hormone at a relatively constant rate out at least to 2 weeks. Thus the Silastic implants appear superior to paraffin implants in maintaining constant plasma estradiol levels.

Figure 7 shows the effect of estradiol on the mass accumulation of total RNA. These data are normalized for DNA content, but the increases are also seen when expressed as mg RNA/uterus or mg RNA/g uterus. An increase is seen at 4 h, the earliest time chosen, and is maximal by about

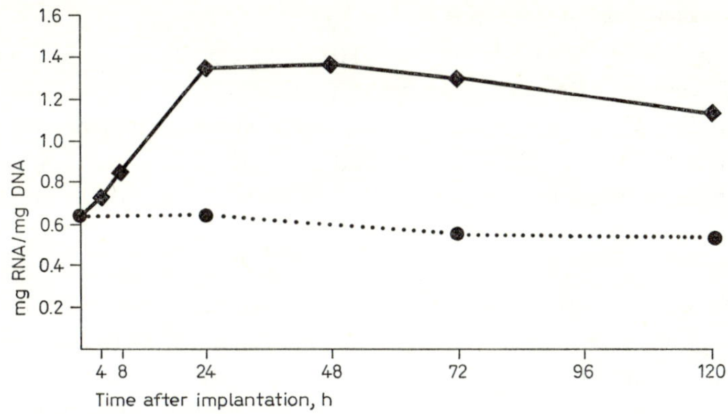

Fig. 7. Effect of estradiol on the uterine RNA/DNA ratio. Ovariectomized rats were sacrificed at intervals after implantation with E_2 in either Silastic capsules or paraffin pellets (♦). Control animals (●) received no E_2. RNA and DNA were determined by the Schmidt-Thannhauser and diphenylamine assays, respectively.

24 h. These observations with implanted animals are consistent with those seen at early times with injected animals [11]. The results (fig. 7) suggest that continuous exposure to estradiol will maintain a constant RNA/DNA ratio and that there are no subsequent major changes in the uterine response. These data, however, do not directly address the question of the molecular consequences of continuous exposure. Several approaches are being used presently to answer this question, including a comprehensive analysis of the poly-A RNA population at different times after hormone administration. Preliminary experiments suggest that the proportion of poly-A containing RNA remains constant at all RNA/DNA ratios. Thus, the doubling of the total RNA/DNA ratio at 24 h is accompanied by a doubling in the poly-A RNA/DNA ratio. Future plans call for an analysis of the sequence complexity of the poly-A RNA after both short- and long-term exposure to estrogen.

Estrogen-Induced Biological Responses in Mouse

As noted in figure 1, the end result of estrogen exposure is to stimulate new protein synthesis at the target cell. This synthesis step underlies any subsequent biological response such as increased uterine weight and perhaps carcinogenesis. Greenman *et al.* [5] have reported on strain differences in the response of the mouse when fed diets containing the estrogen, DES. In this study, BALB/c StCrlfC3Hf/Nctr, C57BL/6/, C57BL/6 × BALB/c F_1 hybrid ($B6CF_1$), and monohybrid-cross offspring from the breeding of $B6CF_1$ mice were examined with respect to uterine, vaginal, and thymus responses to DES. About 400 mice of each generation population were used. Weanling mice were fed DES at dietary concentrations of 2.5–1,000 ppb (μg/kg feed) for 6 days and were killed by cervical dislocation about 20 h after removal of the feed. C57BL/6, $B6CF_1$, and the monohybrid-cross offspring did not differ in the uterine weight response to DES, but the slope of the dose-response line for the BALB/c strain was more shallow than for the other strains. Dietary DES concentrations of 250 ppb or more inhibited the uterotrophic response in all populations. Vaginal cornification occurred at lower concentrations of DES in the C57BL/6 strain than in the $B6CF_1$ animals. BALB/c and monohybrid-cross offspring were indistinguishable from each other in their vaginal response to DES and were less sensitive to DES than the other mouse populations. The use of ethanol or corn oil as the solvent for mixing DES into the diet had no apparent effect on the uterine weight or vaginal response in any of the mice. DES depressed thymus weight in a dose-related fashion at dietary concentrations of $\geq$ 100 ppb in all genetic populations.

Estrogens are carcinogenic in a variety of animal models which raises the basic problem of the potential risk to humans. Long-term studies are underway to evaluate the carcinogenic response of the mouse strains noted above when fed diets containing estrogens but sufficient data has not yet been obtained for statistical analysis in all studies. A concomitant study with the C3H mouse, however, has provided significant results on mammary and reproductive tract estrogen-induced tumorigenesis [6, 7, 12].

The primary objective of this study was to examine further the effects of mouse mammary tumor virus (MTV) factor, DES, or E_2 in estrogen-induced mammary tumorigenesis in female C3H mice. Beginning at 6 weeks of age, high (C3H/HeJ) or low (C3HeB/FeJ) virus-expressing female mice were fed diets containing DES at 0, 10, 100 or 500 ppb and C3H/HeJ female mice were fed E_2 at 100, 1,000 or 5,000 ppb E_2. Each group consisted of 288 animals. At 12 months, the number of mammary tumor-bearing C3H/HeJ mice fed DES were 8 at 0 ppb DES (control), 12 at 10 ppb, 22 at 100 ppb, and 37 at 500 ppb (table I). The number of tumor-bearing C3H/HeJ mice fed E_2 were 7 at 100 ppb, 17 at 1,000 ppb, and 20 at 5,000 ppb. At 18 months, dose-related mammary tumorigenesis in DES-fed mice persisted but was less pronounced in the E_2-fed animals. In the

Table I. Incidence (%) of pathologic changes in mice given diethylstilbestrol (DES) or 17β-estradiol (E_2) for 52 weeks[1]

Agent ppb	Mice, n	Cervix			Uterine horns			Marked ovarian atrophy	Mammary		Osseous hyperplasia
		mucoid stroma	adenosis[2]		glandular hyperplasia	adenomyosis	hyalin changes		HANs	tumors	
			+	>+							
C3H/FeJ (MTV+) mice											
0	47	15	11	0	23	2	4	20	0	4	0
10 DES	32	26	14	0	19	3	3	35	3	0	3
100 DES	38	31	33	6	79	46	39	53	5	8	3
500 DES	48	93	9	89	100	93	85	100	14	7	82
100 E_2	35	17	17	0	24	6	12	15	0	0	0
1,000 E_2	36	34	15	0	62	6	19	29	3	6	16
5,000 E_2	48	100	38	36[3]	96	81	96	91	9	8	82
C3HeB/FeJ (MTV−) mice											
0	18	6	11	0	11	17	11	6	0	0	0
10 DES	39	5	18	0	5	6	3	0	0	0	3
100 DES	18	33	59	0	89	33	50	80	0	0	6
500 DES	37	92	48	21[3]	95	84	95	90	0	0	97

[1] From Highman *et al.* [7].
[2] + Refers to mice with adenosis limited to upper third of cervix and >+ to those with adenosis extending into lower two-thirds.
[3] Significantly different from group on line 4 by χ^2 test, $p < 0.01$. The other groups were not subjected to a statistical analysis.

Table II. Malignant tumors in mice fed diets containing DES or E_2

Strain	Agent	Diet, ppb	Treatment, weeks[1]
C3H/HeJ	DES	500	74[2], 52[4], 52[4], 60[4], 60[4], 68[4], 79[7]
C3H/HeJ	DES	100	75[4], 102[4]
C3H/HeJ	DES	10	78[2], 76[6]
C3H/HeJ	E_2	5,000	52[2, 4], 102[2], 81[3], 66[4], 91[4]
C3H/HeJ	E_2	100	78[4], 58[6]
C3HeB/FeJ	DES	500	52[4], 66[4], 78[5], 52[6]
C3HeB/FeJ	DES	100	93[4]

1 Each number indicates number of weeks a mouse with a malignancy was treated before sacrifice. No such lesions were seen in controls or before 52 weeks. From HIGHMAN *et al.* [7].
2 Adenocarcinoma of uterine horn.
3 Adenoacanthoma of uterine horn.
4 Adenocarcinoma of cervix.
5 Squamous cell carcinoma of vagina.
6 Osteosarcoma of cranium.
7 Osteosarcoma of sternum.

C3HeB/FeJ animals, dietary levels of DES $\leqq$ 500 ppb given for as long as 18 months did not increase the mammary tumor incidence over that observed in the controls. These results reaffirm the roles of DES and the MTV factor in tumorigenesis in female C3H mice and demonstrate that a naturally occurring chemical, E_2, also stimulates mammary tumorigenesis in this animal model system.

Extensive histological examination of these mice revealed marked dose-dependent estrogen-independent reproductive tract lesions. Pathological examinations were made on over 700 mice that were sacrificed or died during the first 15 months. Results are summarized in table I. After prolonged high dosage with either estrogen, the cervix of both populations often showed stromal mucoid changes and adenosis associated with foci of low stratified or columnar epithelium lining the cervical canal. Numerous secretory glands of the endometrial type, frequently showing epidermidization, were often seen, particularly in the upper cervix. The uterine horns showed marked hyperplasia of the glands, which often penetrated the muscularis. Minute secretory glandular inclusions or sprouts were noted focally in the tall epithelium lining the endometrial surface and glands. The ovaries showed atrophy with absence of corpora lutea. Ceroid deposits were increased in the ovaries and adrenals. The sternum showed an increase in bony trabeculae. Hyperplastic alveolar nodules and mammary tumors, mainly type B (Dunn's classification), were frequent in the C3H/HeJ but rare in the C3HeB/FeJ population. To date, a total of 17 adenocarcinomas involving both populations have been observed, variably involving the uterine horns, corpus and cervix (table II). Some appeared to arise from endometrial glands and some from areas of cervical adenosis. Some additional mice showed marked glandular atypia suggesting premalignant uterine lesions. In mice on lower dosages, the findings were similar, but less frequent and severe. These findings indicate that MTV is required for DES- or E_2-induced mammary tumorigenesis in C3H mice, but that other DES- and E_2-induced carcinogenic and precarcinogenic lesions of the reproductive tract may be independent of the MTV.

Overview

Estrogens, in both laboratory animals and man, undergo a series of sequential reactions ultimately leading to an observable biological response in the target tissue. Since the carcinogenic response may be related to the basic mode of action of estrogens, an understanding of these interrelationships is required to fully evaluate the potential risk to humans. The research efforts described in this report are directed to this objective.

Estrogens may be considered carcinogenic on the premise that exposure to an estrogen produces: (a) an increase in incidence of a tumor occurring spontaneously, (b) a decrease in the latent period for tumor induction, or (c) an increased incidence of a rare spontaneous tumor event. But, is this alone a suitable basis for carcinogenic evaluations of hormonally active (specifically estrogenic) chemicals since some estrogens are normally found in all animals? Protocols for evaluating carcinogenicity of estrogens in experimental animals must be reevaluated. Can estrogenic chemicals, even in conjunction with other materials such as progestins, be evaluated in animal models using, by rote, simple multiples of anticipated human exposures? How can results with estrogens be interpreted if the carcinogenic endpoint is complicated by other hormone factors such as prolactin or by viruses? Should toxicological testing protocols be modified to reflect relative potencies of hormonally active agents in the various experimental animals? Answers to these questions will have a significant impact on endocrine toxicology.

However, more important questions also must be resolved. Are estrogens carcinogenic because of estrogenic activity or because of some other unique property akin to mechanisms in chemical carcinogenesis? Are estrogens direct acting carcinogens or functioning in a more complex manner as moderators, i. e. initiators, promotors, or cocarcinogens? How can these latter actions be evaluated? If these events can be titrated, what impact, if any, will they have on regulatory concepts and procedures?

These questions are complex and as yet have not been resolved. The preponderance of data, however, suggest that carcinogenic activity is related to estrogenic potency but a direct permanent action on genetic material cannot be ruled out. It is apparent that a considerable coordinated research effort will be required to answer these questions.

Acknowledgements

The authors thank WILLIAM S. BRANHAM and MARK YOUNG for their technical assistance in these experiments, and J. E. LOWRY for assistance in preparing this manuscript.

References

1 AUSSEL, C.; URIEL, J., and MERCIER-BODARD, C.: Rat α-fetoprotein. Isolation, characterization, and estrogen-binding properties. Biochimie *55:* 1431–1437 1973).

2 BRANHAM, W. S.; CERNOSEK, R. M.; CERNOSEK, S., and SHEEHAN, D. M.: Manuscript in preparation (1977).

3 BRANHAM, W. S. and SHEEHAN, D. M.: Manuscript in preparation (1977).

4 DAUGHADAY, W. H.: Steroid protein interactions. Physiol. Rev. *39:* 885–902 (1959).

5 GREENMAN, D. L.; DOOLEY, K.; BREEDEN, C. R., and GASS, G. H.: Strain differences in the response of the mouse to diethylstilbestrol. J. Toxic. envir. Hlth *3:* 619–627 (1977).

6 HIGHMAN, B.; NORVELL, M. J., and SHELLENBERGER, T. E.: Pathologic changes in mice induced by diethylstilbestrol and 17β-estradiol. Abstract. Toxic. appl. Pharmac. *41:* 178–179 (1977).

7 HIGHMAN, B.; NORVELL, M. J., and SHELLENBERGER, T. E.: Pathologic changes in C3H mice fed continuous diets containing diethylstilbestrol or 17β-estradiol. J. envir. Path. Toxic. (in press, 1977).

8 LAURANT, C.; DELAUZON, S.; CITTANOVA, N.; NUNEZ, E., and JAYLE, M. F.: The comparative specificity of 3 oestradiol-binding proteins. Rat α-fetoprotein, rat liver 17β-hydroxy steroid dehydrogenase and anti-(oestradiol-6-carboxymethyloxime-bovine serum albumin) antiserum. Biochem. J. *151:* 513–518 (1975).

9 LIANG-TANG, L. K. and SOLOFF, M. S.: Characterization of a binding protein

specific for 17β-estradiol and estrone in rat pregnancy plasma. Biochim. biophys. Acta. *263:* 753–763 (1972).
10 McEwen, B. S.; Plapinger, L.; Chaptal, C.; Garlach, J., and Wallach, G.: Role of fetoneonatal estrogen binding proteins in the associations of estrogen with neonatal brain cell nuclear receptors. Brain Res. *96:* 400–406 (1975).
11 Muellar, G. C.; Herranen, A. M., and Jervell, K. F.: Studies on the mechanism of action of estrogens. Recent Prog. Horm. Res. *14:* 95–139 (1968).
12 Norvell, M. J. and Shellenberger, T. E.: Mammary tumorigenesis in female C3H mice fed diets containing diethylstilbestrol or 17β-estradiol. Abstract. Toxic. appl. Pharmac. *41:* 179 (1977).
13 Pearlman, W. H. and Crepy, O.: Steroid-protein interaction with particular reference to testosterone binding by human serum. J. biol. Chem. *242:* 182–189 (1967).
14 Raynaud, J. P.; Mercier-Bodard, C., and Baulieu, E. E.: Rat estradiol binding plasma protein (EBP). Steroids *18:* (767–788 (1971).
15 Sandberg, A. A.; Slaunwhite, W. R., and Antoniades, H. N.: Binding of steroids and steroid conjugates to human plasma proteins; in Pincus Rec. Progr. in Hormone Res., vol. 13, p. 209–260 (Academic Press, New York 1957).
16 Sanu, L.; Nunez, E., and Jayle, M. F.: Plasma diethylstilboestrol binding proteins of rat, mouse and man in the course of development: relations with the binding of oestradiol. Steroids *25:* 717–728 (1975).
17 Sheehan, D. M.: Manuscript in preparation (1977).
18 Soloff, M. S.; Creange, J. E., and Potts, G. O.: Unique estrogen-binding properties of rat pregnancy plasma. Endocrinology *88:* 427–432 (1971).
19 Soloff, M. S.; Morrison, M. J., and Swartz, T. L.: A comparison of the estrone-estradiol-binding proteins in the plasmas of prepubertal and pregnant rats. Steroids *20:* 597–608 (1972).
20 Swartz, S. K.; Soloff, M. S., and Suriano, J. R.: Binding of estrogens by α-fetoprotein in rat amniotic fluid. Biochim. biophys. Acta *338:* 480–488 (1974).
21 Young, M. and Sheehan, D. M.: Manuscript in preparation (1977).

Dr. T. E. Shellenberger, National Center for Toxicological Research, Building 15, *Jefferson, AR 72079* (USA)

Discussion

Van Wimersma Greidanus: Do you expect your binding studies in plasma to provide answers to our questions regarding the interactions of hormones? Are the receptors in different tissues all identical and, if so, may we expect the same kinds of hormone interaction?

Shellenberger: In our blood binding studies, the differential binding of estradiol and DES to the α-fetoprotein (α-FP) in the pregnant plasma were rather surprising. I do not know if there is a different receptor activity in the various tissues. I wonder if the blood binding cannot be modeled as a means to carry the active estrogen to the target cell.

VAN WIMERSMA GREIDANUS: If you are correct in assuming that there is a carrier system in the blood, some, if not all, aspects of this might be applied to the intercellular events where a carrier system exists from the outside to the inside of the nucleus.

SHELLENBERGER: You may be right, but I think that at this stage we can only speculate.

VOIGT: As far as blood studies are concerned, rat plasma does not contain the sex hormone binding globulin, so you cannot apply findings in the rat to the human situations.

SHELLENBERGER: The rat α-FP or the high affinity binder in the rat plasma may be analogous to human testosterone-estrogen binding globulins, but we have not done enough work on that.

VOIGT: LASNITZKI and FRANKLIN [J. Endocr. *64:* 289 (1975)] have demonstrated that there is a correlation between steroid binding to SHBG and the *in vitro* uptake of testosterone by cell cultures of prostatic tissue. On the other hand, there is the situation where you have no SHBG and no disturbances in steroid uptake and metabolism. I doubt that these proteins have any meaning under biological conditions.

SHELLENBERGER: I share your doubts. We started our work with the pregnancy plasma because of the alleged transplacental effects of DES on girls whose mothers had been treated. We do know that E_2 and DES pass the placenta in rodents and we wanted to study the carrier mechanism. We hope that if we can continue this work we shall be able to use it as some indication of relative activity of estrogens. A lot of work remains to be done, but we have shown already that, particularly with regard to its behaviour to α-FP, DES behaves differently from E_2.

BAULIEU: There are considerable differences in affinities of steroids to α-FP. Estrone is bound more than estradiol, estradiol more than DES. The SBP in the adult human, who has no α-FP, binds estradiol, but hardly any estrone or DES. We now have antibodies against both proteins and we hope that, before long, we shall have more insight into their distribution and function. DES is an effective estrogen, as good as estradiol, and the fact that there are considerable differences in binding to α-FP and to SBP indicates that the effectiveness cannot be just dependent on a carrier system. You provoked uterine tumours in the C3H mice after loading them with estrogen. Were these tumours, later on, still estrogen-dependent or estrogen-sensitive?

SHELLENBERGER: We do not know yet. Our experiments to clarify this are still in progress.

BAULIEU: I asked this because of the observations of NOBLE [NOBLE *et al.:* Cancer Res. *37:* 1455–1460 (1977)] who found that while massive doses of estrogens provoked tumours in rats, these tumours became estrogen insensitive when the animals were subsequently kept on a low estrogen intake. It seems that estrogens may well be beneficial at one level, and dangerous at another, much higher one.

SHELLENBERGER: Your point is well taken. Of course, the toxicologist tries to increase the sensitivity of the animal model when he has to do a carcinogenic bioassay, because he is limited in the number of animals that he can use. The usual way to do this is by administering massive doses. The selection of the maximum tolerated dose (MTD) is usually based on growth curves, weight depression or food consumption data, but when studying hormones, and estrogens in particular, this probably is a not very adequate method. In this case, we should perhaps select the MTD after studying the effects of the exogenous estrogen on the endogenous levels. However, if we seek to change the concept of endocrine toxicology, we have also to change the concepts of those who review the NDA and IND. Let me give another example: oral contraceptives with a progestagen:estrogen ratio of 10:1 are tested in rodents in simple multiples of the dose in clinical use, e. g., the dose × 1, × 10 and × 100. In the rat the progestagen levels are then not physiologically active, but the estrogens are hyperphysiological. You are then in fact testing the estrogen, but not the progestagen, and certainly not the combination. The thing to do would be to alter the ratio for such a study in rats but then, the reviewers say, the results cannot be applied to the product marketed for humans. This is the dichotomy of the review regulatory process for chemicals such as these, possessing physiological activity.

BAULIEU: Coming back to the receptors, preliminary results indicate that the steroid receptors in the various tissues, including breast cancer tissue, are similar and it is not excluded that they are the same. There are also cross-immunoreactions between estradiol receptors of certain mammals.

VOOIJS: How many malignant tumours of the cervix were induced with DES in these animals?

SHELLENBERGER: At the time this report was made there were 17, but the figures are, of course, subject to updating. There had however been none in the control animals, and that is what makes our findings, even at this stage, so distinctive.

VOOIJS: The morphology of the tumours you have shown seems to be different from the morphology of DES-linked tumours in the human.

SHELLENBERGER: That may be, but for the interpretation I have to rely on others. We are certainly getting uterine cervix changes and I was surprised by this. I had read about the C3H mouse and mammary tumorigenesis, but never found a description of uterine cervical changes in feeding studies. I presume earlier investigators did not look at that tissue.

Front. Hormone Res., vol. 5, pp. 220–229 (Karger, Basel 1978)

On the Biological Activity of Estrone *in vivo*

J. H. H. THIJSSEN, M. A. H. M. WIEGERINCK[1], G. MULDER and J. POORTMAN

Department of Endocrinology, and [1]Department of Obstetrics and Gynecology, University Hospital, Utrecht

Introduction

During the previous workshop [1, 2] the mechanism of estrogen production in postmenopausal women by the peripheral conversion of androstenedione to estrone was discussed in detail. Whereas estradiol is the most abundant estrogen in the blood of women during their reproductive period, the main circulating estrogen in postmenopausal women becomes estrone. A specific effect of estrone, different from estradiol, on the target cells has been postulated by SIITERI *et al.* [3]. In this so-called 'estrone hypothesis' they have suggested that unopposed exposure of target tissues to estrone may be a causal factor in the development of cancer. In essence, they propose that endogenously produced estrone is carcinogenic, whereas estradiol and estriol are not.

Some *in vitro* experiments, however, have shown that even when only estrone is supplied to human endometrium, the estrogen tightly bound to nuclei is not estrone but estradiol, formed from estrone in the tissue [4]. It is therefore uncertain whether in humans estrone is estrogenic *per se* or becomes estrogenic by conversion to estradiol at the target tissue as well as peripherally.

Therefore, a study was initiated to measure tissue estrogens *in vivo* in postmenopausal women after administration of estrone via a continuous infusion. For this infusion a time of 12 h was chosen because the time needed to reach an equilibrium in the reaction estrone $\rightleftarrows$ estradiol *in vivo,* measured in peripheral blood, takes several hours [5, 6].

In this paper the preliminary results, obtained in 6 patients will be described.

Experimental Design

All patients participated after informed consent. 6 women who were scheduled to undergo a hysterectomy for non-oncological reasons, received a constant infusion with ^{3}H-estrone, starting about 12 h before the operation. Some information on these patients is given in tables I and II. All had normal liver and kidney function, no endocrine abnormalities like thyroid disorders or diabetes were present.

At the time of operation, peripheral blood samples were taken to mea-

Table I. Summary of some data of the patients studied: all women were without hormonal therapy for at least 6 months, only patients 4 (Ek) received medication (Hydergine® and Baycaran®) at the time of the operation

Patient	Age years	Age at menopause years	Weight kg	Height m	Blood pressure	Indication for operation	Histological findings of uterus
1 Sc	53	47	50	1.65	130/95	prolaps of the uterus, urinary stress incontinence	cystic glandular hyperplasia of endometrium, normal myometrium
2 Qu	51	47	62	1.60	120/80	chronic low abdominal pain unilateral ovarian cyst	normal endometrium normal myometrium
3 De	55	50	58	1.55	130/100	uterine myoma cystocele	normal endometrium myoma
4 Ek	53	51	64	1.64	130/85	uterine myoma postmenopausal bleeding	slight cystic glandular hyperplasia of endometrium myoma
5 We	58	49	93	1.63	170/105	recurrent postmenopausal bleeding	cystic glandular hyperplasia of endometrium
6 Bo	57	55	105	1.69	160/105	recurrent postmenopausal bleeding	normal endometrium

Table II. Some hormone levels in plasma of the patients studied: in patient 4 (Ek) the estrogen and the gonadotrophin levels are not consistent with post-menopausal status

Patient	FSH mIU/ml MRC 68/39	LH mIU/ml MRC 68/40	Estrone ng/l	Estradiol ng/l	Testo-sterone µg/l
1 Sc	–	–	70	10	–
2 Qu	47	43	22	8	0.13
3 De	–	64	26	7	0.23
4 Ek	9	10	76	83	0.26
5 We	–	78	89	23	0.62
6 Bo	31	31	–	–	0.24

sure the concentration of ^{3}H-estrone and of ^{3}H-estradiol in the plasma, using methods previously described [7]. Usually three samples were collected, at the beginning of the operation, 30 min and approximately 60 min later. The last sample was taken at the time of the removal of the uterus. After removal and immediate dissection of the tissue by the pathologist, pieces of myometrium and of endometrium plus adjacent myometrium were either stored at –70 °C or processed immediately.

The infusion rate of ^{3}H-estrone in each patient was approximately 4 µCi/h, amounting to a total dose of 50 µCi/patient. At the specific activity of ^{3}H-estrone of 50 Ci/mmol, this dose equals 0.3 µg of estrone, i. e., less than 2 % of the endogenously produced estrone during the experimental period of 12 h. The total radiation dose due to the infusion experiments was calculated for the patients to less than 2 mrem.

In the tissues obtained from patients 1 to 4 the level of protein-bound radioactive estrogens was measured by the method given in figure 1. The homogenization of the tissues was done in a buffer containing 2.5 ng of estrone and of estradiol per gram wet weight of the tissue. Addition of the nonradioactive estrogens was done to prevent binding of radioactive estrogens, present in the blood, to nonoccupied (free) estrogen receptor sites in the tissues during the homogenization procedure.

Due to technical problems with the infusion pump, the ^{3}H-estrone infusion in patient We (No. 5) was stopped about 2 h before removal of the uterus. Tissues were processed as for the first 4 patients (fig. 1).

The tissues obtained from the last patient (Bo) were processed fresh, immediately following removal of the uterus, by the method given in figure 2. Special care was taken to get a preparation of pure nuclei, by using the method described by DE BOER *et al.* [8].

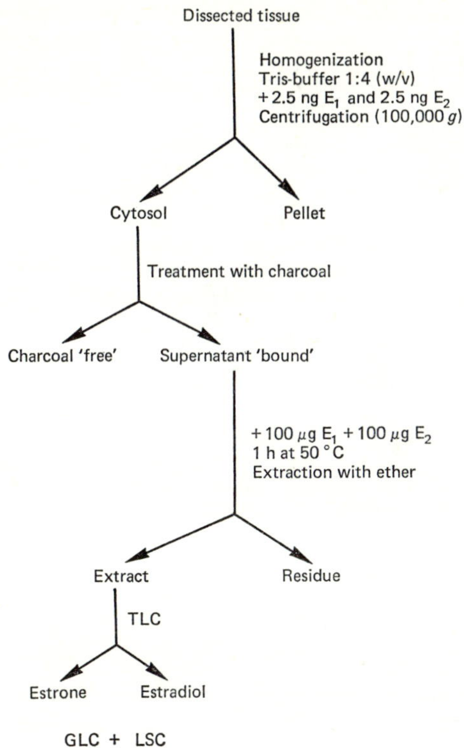

Fig. 1. Method used to measure and characterize the protein-bound estrogens in the uterus of postmenopausal women after infusion of ^{3}H-estrone. E_1 = estrone; E_2 = estradiol; TLC = thin-layer chromatography; GLC = gas-liquid chromatography; LSC = liquid scintillation counting.

As indicated in both figures, during the last stages of processing, non-radioactive estrone and estradiol were added to the fractions to be identified. In this way the losses occurring during the isolation and chromatographic purification could be monitored. All quantitative data given in the following section have been corrected for these losses.

Results

In the myometrium and endometrium of the first 5 patients, the level of radioactive estrone and estradiol was estimated, bound to protein in the cytosol after charcoal treatment. This treatment does not distinguish be-

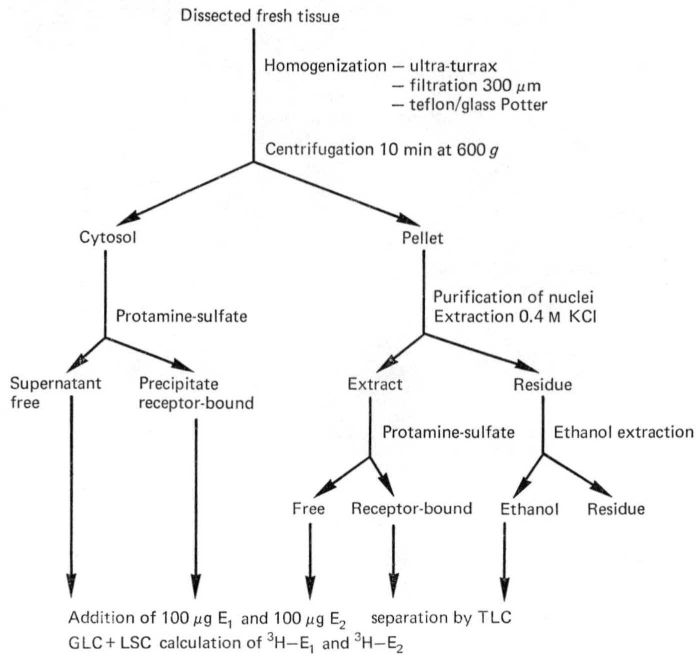

Fig. 2. Method used to identify the receptor-bound estrogens in the endometrium and myometrium of postmenopausal patient Bo after infusion of ^{3}H-estrone. Both tissues were processed directly after removal of the uterus. For abbreviations see figure 1.

tween specific receptor binding and binding to possibly contaminating proteins with a relatively high binding affinity. In the sixth patient, the level of receptor-bound estrogens in the nucleus and the amount of nonreceptor-bound estrogens in both the cytosol and the nuclei could also be obtained. In this patient the protein binding was further characterized by the specific protamine-sulfate precipitation of the receptor-complex [9]. In table III a summary of the results is given as the ratio between ^{3}H-estradiol to ^{3}H-estrone in the plasma and in the tissues studied.

From these figures it is evident that the concentration of ^{3}H-estradiol, compared to ^{3}H-estrone found protein-bound in the cytosol of the tissues in patients 1, 2 and 3 is much higher in these tissues than in plasma. Even when the different affinity of estradiol and of estrone towards the receptor is taken into account, the relative increase in ^{3}H-estradiol bound to protein

Table III. The ratios of ^{3}H-estradiol to ^{3}H-estrone measured in the plasma and in some tissues of postmenopausal women after infusion of ^{3}H-estrone

Patient	Plasma[1]	Cytosol myometrium protein-bound[2]	Cytosol endometrium protein-bound[2]
1 Sc	< 0.20	0.7	1.0
2 Qu	0.10	0.7	0.9
3 De	0.13	9.3	3.6
4 Ek	< 0.30	0.4	–[4]
5 We	0.14	11.0	1.7
6 Bo[3]	< 0.20	1.5[5]	0.5[5]
		0.4 (free)	–[4]
		nucleus	nucleus
		> 1[5]	> 1[5]
		0.1 (free)	0.1 (free)

[1] Due to inaccuracy in the measurement of the ^{3}H-estradiol levels in plasma, the exact figures cannot be given, the figures are the maximal estimates of the ratio.

[2] For patients 1–5 protein-bound indicates the ratio calculated after charcoal treatment of the cytosol.

[3] In this patient the protein-bound estrogens were characterized by specific protamine-sulfate precipitation of the receptor complex. The nonprecipitated radioactivity could therefore also be identified (free).

[4] Not estimated.

[5] Characterization by protamine-sulfate precipitation, no significant amounts of estrone were found in the nucleus.

('receptor') is clear. This increase was not found in the myometrium of patient 4. This woman, however, was the only one with premenopausal estrogen concentrations in blood; her estradiol level in plasma was 83 pg/ml. This level indicates a production of estradiol separate from the conversion of estrone to estradiol.

The increase in estradiol concentration of the receptor-bound radioactivity is even more evident in patient 5, in whom the infusion was stopped 2 h before removal of the uterus. In accordance with animal experiments the suggestion is clear that the retention of estradiol in estrogen-sensitive target cells is longer than that of estrone.

Because all these tissues had been stored at –70 °C during periods of up to several weeks, no nuclear uptake studies could be done, due to the breaking down of the nuclei during freezing and thawing.

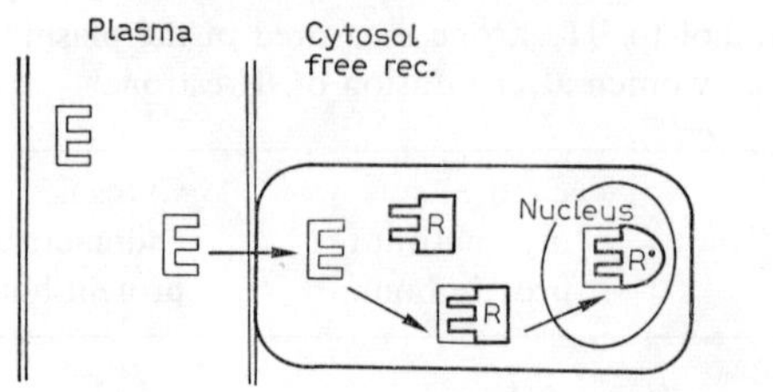

Fig. 3. The ^{3}H-estradiol/^{3}H-estrone ratio in peripheral plasma and in the myometrium of patient 6 is shown after infusion of ^{3}H-estrone for 12 h. In the tissue the receptor-bound estrogen ratio is given and the ratio of the free estrogens. Receptor binding was characterized by specific precipitation of the estrogen receptor complex with protamine sulfate.

Therefore, in the last patient studied so far, care was taken to isolate the nuclei of the tissues investigated immediately after removal of the uterus. From the ratios given in table III for patient 6, it is clear that the previously obtained results are being confirmed by the observed ratio in the nuclear-bound fractions. In addition, the proteins to which the estrogens are bound were characterized as the receptor by the protamine-sulfate precipitation. For the myometrium of this patient, all results are shown in figure 3.

Discussion and Conclusions

The levels of radioactive estradiol and estrone in peripheral plasma, measured between 11 and 12 h after the beginning of a constant infusion of ^{3}H-estrone, indicated that the equilibrium in the conversion of estradiol $\rightleftarrows$ estrone had been reached. Of course, it is still uncertain whether the equilibrium had also been reached in the tissues studied.

The *first conclusion* that can be drawn, from the differences in the ratio of ^{3}H-estrone to ^{3}H-estradiol in the tissues (protein-bound) compared to plasma, is that the estrogen ratio in the uterus does not reflect this ratio in peripheral blood. The tissue ratio is shifted towards more estradiol.

To explain this relative increase in estradiol concentration, several possibilities can be considered:

(a) The ratio between estrone and estradiol in blood depends partly on the *metabolic clearance rate* of each compound. Data available in postmenopausal women do not show a large difference in the metabolic clearance rates. Whether there are differences in the clearance rate of each compound in uterine tissue is not known.

(b) A *selective uptake* of estradiol by the uterine cells from the plasma. Direct evidence is neither available for a selective carrier mechanism for steroids across membranes nor for differences between estrone and estradiol in this respect. A higher uptake of free estradiol from plasma by the cells is not supported by the higher free (nonprotein-bound) estrone concentration in the plasma, because of estradiol binding to sex hormone binding globulin.

(c) A higher *relative binding affinity* of the specific estrogen receptor in the cells. This factor could explain a maximally twofold increase in estradiol concentration because of its higher binding affinity as compared to estrone.

(d) *Local factors* such as intracellularly enzymatic activity. It has been shown that uterine cells contain the 17β-hydroxysteroid oxidoreductase. The activity of this enzyme, important for the conversion of estrone to estradiol, is influenced by progestational compounds [10].

(e) A *longer retention* of the estradiol receptor complex in the cells. This possibility is supported by results from experiments with animals [11] and by our results in patient 5 (table III). In this patient the infusion with ^{3}H-estrone was stopped 2 h before removal of the uterus, resulting in a very high ^{3}H-estradiol/^{3}H-estrone ratio in the tissue.

In our opinion, local factors favoring the formation of estradiol from estrone in the cells play a major role for the explanation of our findings. The resulting higher intracellular concentration of estradiol will result in increased binding to the available receptor sites in the cytoplasm.

Because of the known influence of progestational compounds on the 17β-hydroxysteroid oxidoreductase activity, our data obtained in patient 4 may be explained by this factor. In this obviously premenopausal woman no progesterone determinations are available. Therefore we cannot exclude the possibility that the enzymatic activity might also be influenced by the ratio of the levels of estradiol and estrone in peripheral plasma.

Our preliminary data show a consistently higher estradiol/estrone ratio in uterine tissue when compared with peripheral plasma. This altered ratio in favor of estradiol is even more pronounced in the specific protein-bound fraction in the cytoplasm and in the nuclei of the uterus. This accumulation of estradiol makes the estrone hypothesis in its original form as presented by SIITERI *et al.* [1, 3] highly unlikely. On basis of our findings it is more likely that the relation between certain malignancies and an increased estrone production cannot be explained by specific separate effects of estrone but by effects due to an increased biological activity of estrogens.

References

1 SIITERI, P. K.: Postmenopausal estrogen production; in VAN KEEP and LAURITZEN Estrogens in the post-menopause. Front. Hormone Res., vol. 3, pp. 40–44 (Karger, Basel 1975).

2 THIJSSEN, J. H. H.; POORTMAN, J.; SCHWARZ, F., and DE WAARD, F.: Post-menopausal estrogen production, with special reference to patients with mammary carcinoma; in VAN KEEP and LAURITZEN Estrogens in the post-menopause. Front. Hormone Res., vol. 3, pp. 45–62 (Karger, Basel 1975).

3 SIITERI, P. K.; SCHWARZ, B., and MACDONALD, P. C.: Estrogen receptors and the estrone hypothesis in relation to endometrial and breast cancer. Gynecol. Oncol. *2:* 228–238 (1974).

4 TSENG, L. and GURPIDE, E.: Effect of estrone and progesterone on the nuclear uptake of estradiol by slices of human endometrium. Endocrinology *93:* 245–248 (1973).

5 HEMBREE, W. C.; BARDIN, C. W., and LIPSETT, M. B.: A study of estrogen metabolic clearance rates and transfer factors. J. clin. Invest. *48:* 1809–1819 (1969).

6 LONGCOPE, C. and TAIT, J. F.: Validity of metabolic clearance and interconversion rates of estrone and 17β-estradiol in normal adults. J. clin. Endocr. Metab. *32:* 481–490 (1971).

7 POORTMAN, J.; THIJSSEN, J. H. H., and SCHWARZ, F.: Androgen production and conversion to estrogens in normal postmenopausal women and in selected breast cancer patients. J. clin. Endocr. Metab. *37:* 101–109 (1973).

8 BOER, W. DE; VRIES, J. DE; MULDER, E., and VAN DER MOLEN, H. J.: Comparative study of nuclear binding sites for oestradiol in rat testicular and uterine tissue. Determination of low amounts of specific binding sites by an ^{3}H-oestradiol-exchange method. Biochem. J. *162:* 331–339 (1977).

9 STEGGLES, A. W. and KING, R. J. B.: The use of protamine to study 6,7-^{3}H-oestradiol-17β binding in rat uterus. Biochem. J. *118:* 695–701 (1970).

10 GURPIDE, E.; GUSBERG, S. B., and TSENG, L.: Estradiol binding and metabolism in human endometrial hyperplasia and adenocarcinoma. J. Steroid Biochem. *6:* 891–896 (1976).

11 JENSEN, E. V. and JACOBSON, H. I.: Basic guides to the mechanism of estrogen action. Recent Prog. Horm. Res. *18:* 387–414 (1962).

Dr. J. H. H. THIJSSEN, Department of Endocrinology, University Hospital, Catharijnesingel 101, *Utrecht* (The Netherlands)

Discussion

LAURITZEN: Your findings – preliminary as they may be – force us to re-appraise the hypothesis of SIITERI and MACDONALD presented by SIITERI at our last meeting [Estrogens in the post-menopause. Front. Hormone Res., vol. *3,* pp. 40–44 (Karger, Basel 1975)].

BAULIEU: These *in vivo* findings of Dr. THIJSSEN are certainly compatible with experimental models. There seems to be a selective 'trapping' of estrogens by the receptor, whenever estradiol and estrone are offered, according to their respective affinity and concentration. I too believe that there is little evidence that the estradiol receptor complex has properties different from those of the estrone receptor complex, apart from the fact that the latter is more labile.

THIJSSEN: SIITERI misquoted the work of GURPIDE at our last meeting. GURPIDE did *in vitro* perfusions of human endometrium with estrone and, according to SIITERI, found that this was bound to the receptor, but this was not so.

BAULIEU: A piece of tissue, even when superperfused, is 'dying' and this has many consequences: the enzymatic activities, including that of 17β-dehydrogenase may well be different from what you observe *in vivo.*

ROGER KING of London has been working with conjugated equine estrogens and has studied the estrogen receptor at the human endometrial level. He has recently found that when a daily dose of 0.625 mg or 1.25 mg of Premarin was given – not an excessively high one in clinical practice – the amount of estrogen found at the receptor level was as high as that found under physiological circumstances during the fertile age [KING, R. J. B.; WHITEHEAD, M.; CAMPBELL, S., and MINARDI, J.: Biochemical studies on endometrium from postmenopausal women receiving hormone replacement therapy. Post-grad. med. J., in press 1978].

VOIGT: Dr. THIJSSEN, how did you cope with plasma contamination? The binding protein in plasma binds estradiol more than estrone, and this might have distorted your findings.

THIJSSEN: We tried to quantify the amount of blood in the homogenates by measuring haemoglobin in the final cytosol, and found that there was little contamination. In the last experiment we did a protamine-sulphate precipitation to be more sure about the receptor binding.

BAULIEU: We have also found that there is not a significant amount of sex steroid binding plasma protein in endometrium extract [MERCIER-BODARD, C.; ALFSEN, A., and BAULIEU, E. E.: Acta endocr. Copenh., *64:* suppl. 147, pp. 204–244 (1970)].

Front. Hormone Res., vol. 5, pp. 230–247 (Karger, Basel 1978)

Management of the Patient at Risk

C. LAURITZEN
Department of Gynaecology and Obstetrics, University of Ulm, Ulm

Medicaments which are highly effective cannot act without exerting side effects. For some patients certain of them may even involve actual risk. In order to minimize the known and to some degree foreseeable dangers of a medication, respect must be paid to indications and contra-indications. The contra-indications for estrogen therapy in the post-menopause are listed in table I. It is, of course, not sufficient for the prescribing physician to know when hormone treatment should and should not be given; he must also know which symptoms warrant its withdrawal once prescribed (table II).

With all therapies one has to think in terms of the risk factors and patient risk groups in order, it is hoped, to be able to prevent as many as possible of the undesirable or even dangerous side effects. As far as estrogen therapy is concerned no studies have yet been published demonstrating that, in this way, any real reduction in the number of side effects or risks can be attained. Thus, for the time being, consideration of the risks and of the indications for treating risk groups must be based on personal experience, on a comparison with other available treatments, and on deductive conclusions from epidemiological statistics.

Most of our knowledge concerning the side effects of estrogens is drawn from experience with oral contraceptives which contain artificial C-17-alkylated estrogenic compounds. This has, regrettably, led to many generalizations concerning 'estrogens'. It must be pointed out that when treating climacteric complaints natural estrogens are usually used, and with these far fewer side effects have been experienced than with the estrogen-progestagen hormonal compounds of the contraceptive pill. This may be due to the fact that much lower equivalent doses of estrogens are

Table I. Contra-indications for estrogen therapy

Absolute
Severe, acute liver diseases and severe liver damage due to chronic disease
Enzymopathies of the liver
Porphyria
Acute gallbladder and pancreas diseases
Sickle cell anaemia
Deep venous thrombosis, thrombo-embolism
Cerebro-vascular diseases and malformations
Estrogen-dependent tumours still under supervision (endometrial cancer, mammary cancer, malignant melanoma)
Relative
Myomata[1], endometriosis[1]
Mastopathy[1]
Endometrial and mammary carcinoma, probably cured[1]
Pituitary tumour, hyperprolactinaemia
Severe diabetes with vessel degeneration
Fixed hypertension
Hyperlipidaemia
Oedema, heart diseases, renal diseases
Epilepsy, otosclerosis
Severe migraine
Hyperpigmentation
Diseases which are adversely affected by pregnancy
Impending operations, immobilization
High dose corticosteroid medication
Proven overstimulation of the endometrium

[1] In these cases only low doses of estrogen should be given, and the addition of a progestagen is recommended. The use of estriol should be considered. Close medical supervision is necessary.

used than in the contraceptive pill. A dose of about 1.25 mg of conjugated estrogens, or of 2 mg micronized estradiol or estradiol valerate is equivalent to about 25 μg of ethinyl estradiol as far as all important parameters are concerned. It seems that the occurrence of severe side effects, such as thrombo-embolism, is dose dependent [Inman *et al.*, 1970].

It is also felt that there may be differences in the effects and in the side effects of the various estrogens, differences between, for instance, the artificial estrogens (stilbestrol, mestranol, and ethinyl estradiol) and the estrogens occurring naturally in the human (estrone, estradiol-17β, and

Table II. Reasons for the withdrawal of estrogens

Essential
Icterus, gallbladder pain, pancreatitis
Increase of liver enzymes
Acute porphyria
Deep venous thrombosis, embolus
Myocardial infarction, cardiac insufficiency
Cerebro-vascular accident
Severe migraine, visual disturbances
Rapid growth of pigment spots (melanoma)
Rapid deterioration of hearing (otosclerosis)
Epileptic seizures
Impending operations
Mammary or genital cancer
Recommended
Long immobilization
Oedema, rapid weight gain
Increase of blood pressure
Allergic skin reaction (to estrogen)
Rapid growth of myoma, endometriosis and breast nodule
Hyperplasia of endometrium, undiagnosed genital bleeding
High dose corticosteroid medication

estriol), and in equines (conjugated equine estrogens). Differences in the effects on the endometrium (estriol, HEUSER and STAEMMLER [1973]), on blood coagulation [DAVIES *et al.*, 1975], and on lipid fractions [LEBECH and BOGGAARD, 1974] have been described, but the significance of quantitative differences and of the importance of different dose equivalents is not yet completely clear.

It is obvious that when reporting on the use and effects of 'estrogens' the compounds and the dose ranges should be clearly stated.

In my opinion the use of artificial estrogens, at least in doses higher than the equivalent of 25 μg ethinyl estradiol, is best avoided when treating the climacteric woman. Mestranol and ethinyl estradiol have a high rate of side effects and are strongly endometriotrophic, which is certainly not appreciated in the post-menopause. These hormones, especially when given orally, stimulate directly and very intensively and undesirably, many biosynthetic reactions in the liver, and are more cholestatic than steroids which are not C-17-alkylated. They are strongly bound to the receptors and have a long half-life time (36–40 h) in plasma against the short one (3–6 h) of estrone, estradiol and estriol [KOLB, 1967; BOLT and BOLT, 1974; LONGCOPE and WILLIAMS, 1974; FOTHERBY, 1976; ENGLUND and JOHANSSON, 1977].

Therefore, when a daily medication of ethinyl estradiol is given, cumulation and consequently overstimulation of the genital target organs and breasts are to be expected after some weeks. In addition the 17α-ethinyl group cannot be removed by the organism. Therefore, ethinyl estradiol is not metabolized to an impeded metabolite like estriol, which normally seems to modify the biological effects of the primary estrogens estrone-estradiol at the target organs [VELARDO, 1958; LAURITZEN, 1962]. Endometrial overstimulation by estrogens is apparently less critical when a progestagen is added.

Potential Dangerous Side Effects

Only the potential dangerous side effects of estrogen medication will be dealt with here. These can be: increase in blood pressure, decrease of

Table III. Factors which have to be respected in anamnesis

Family history
Cancer, diabetes, lipid diseases, vascular diseases, gallbladder diseases
Individual history
Cancer, other tumours
Cycle anomalies, curettages
Breast tumours, mastopathy
Pregnancy histories (glucose anomalies, liver disease, hypertension, gestosis)
Thrombosis, embolus
Sickle cell anaemia
Angina pectoris, heart and coronary diseases, myocardial infarction
Cerebro-vascular accidents
Diabetes, hyperlipidaemia, hypertension
Liver and gallbladder disease, pancreatitis
Smoking habits, alcohol use
Operations
Intake of medicaments (tranquillizers, corticosteroids)
Side effects during oral contraceptive use

Table IV. Factors which have to be respected in physical examination

Overweight
Hirsutism
Myomata
Endometriosis
Hysterectomy, oophorectomy, ovarian tumour
Mastopathy
Galactorrhoea
Liver and gallbladder scars
Hypertension
Glucose/protein in urine
If the anamnesis or physical examination warrants it, the following tests should be carried out:
Determination of cholesterol, triglycerides, transaminases, alkaline phosphatase, leucinamino-peptidase, BSP test

glucose tolerance, elevation of triglycerides in plasma, cholestasis, increased risk of cholecystitis, cholelithiasis and pancreatitis, and the growth of myomata and endometriosis. Potentially life-threatening are the deep vein thrombosis with the risk of an embolus, myocardial infarction, cerebral vascular accident, and the possibility of a promotion of an endometrial or mammary carcinoma. Tables III and IV depict the factors which have to be respected in anamnesis and physical examination when considering treating a patient with estrogens.

Hypertension

While, according to several investigators, oral contraceptives increase the blood pressure in 3–5 % of all users, the percentage is apparently lower when the estrogens administered are natural ones. Several papers state that the long-term administration of estrogens, when estradiol, conjugated estrogens, or estriol have been used, has not been found to increase mean or individual blood pressures [DAXELMUELLER *et al.*, 1974; RAURAMO, 1975; BORGLIN and STALAND, 1975; NOTELOVITZ, 1975]. In cases where an increase has been seen when rather high doses of conjugated estrogens were used [UTIAN, 1975], the role of estrogens is not clear; it may be that na-

Table V. Risk factors for hypertension

Positive family history
Renal diseases
Earlier hypertensive episodes (during pill intake, pregnancy, EPH gestosis)
High salt intake
Overweight, diabetes
Plethora
Hyperlipidaemia
Aortal stenosis and insufficiency
Hyperthyroidism
Acromegaly
Cushing syndrome
Conn syndrome (hyperaldosteronism, porphyria)

tural estrogens have been causal as most of the women involved had either hypertension in the family anamnesis or severe renal diseases. The cause of hypertension in these latter cases is a disturbance in the renin-angiotensin-aldosterone system because of a feedback error in the inhibition of renin substrate. The aldosterone level is usually elevated [NEWTON *et al.*, 1968; CRANE *et al.*, 1971]. It is clear, therefore, that the blood pressure of the patient should be recorded before estrogen therapy is prescribed.

The risk factors for the development of hypertension are listed in table V. When any of these is present there is a relative contra-indication for the administration of estrogens. It has been claimed, however, that the use of a low dose of natural estrogens and the reduction of salt intake to below 3 g/day may prevent the occurrence of hypertension [VON EIFF, 1975].

When an increase of blood pressure occurs under therapy, a regression is sometimes achieved by changing the preparation or by lowering the dose. In most cases estrogen should be withdrawn when hypertension develops. Sometimes, however, one can additionally give anti-hypertensives for a while. Under such treatment the blood pressure usually normalizes and, if no other diseases are present, sometimes remains stable even after the withdrawal of the anti-hypertensive medication.

It has also been claimed that progestagens play a role in the development of hypertension [Royal College of General Practitioners, 1974]. If this is suspected the progestagen should be changed, and only withdrawn if the problem then persists.

Decrease of Glucose Tolerance

During the intake of some oral estrogens a decrease of glucose tolerance and of the K-value has been registered in about 30% of cases. The oral glucose tolerance test is said to give abnormal results more often than the intravenous test. The cause of the deterioration of glucose tolerance is possibly an increase of growth hormone secretion during estrogen treatment and a change of cortisol concentration and binding in plasma [SPELLACY and BIRK, 1972; KALKOFF, 1972; PYÖRÄLÄ, 1976; SYVALAHTI *et al.*, 1976].

An influence on intestinal hormones has also been suspected. Apparently oral estriol or estriol succinate do not influence the oral glucose tolerance test [STROHFELDT, 1977]. Estrogens and progestagens have no direct diabetogenic effect. Manifest diabetes is not an absolute contra-indication against the intake of estrogens and progestagens unless severe vascular alterations are present. It is desirable, however, to control more exactly the metabolism of those pre-diabetic or diabetic patients, who receive estrogens. Often an adaptation of the diet and a diminution of the estrogen dose is sufficient for normalization. Progestagens influence the glucose tolerance much less than estrogens do. In some cases the glucose tolerance and K-values are normalized naturally after a few months of treatment. After withdrawal of the estrogens a normalization is achieved within a few weeks. Since the patients who react with a decrease of glucose tolerance are often latent diabetic, pre-diabetic, adipose, or have a familial diabetic predisposition, the family history in this connection should always be elucidated before estrogens are prescribed.

Disturbance of Lipid Metabolism

Artificial and natural estrogens, such as estradiol and estrone, cause a decrease of total lipids. In particular, β-lipoprotein cholesterol (low density lipoprotein), which is quantitatively the most important fraction, is lowered [FURMAN, 1969; BOYD, 1973]. This effect is considered advantageous. While ethinyl estradiol and mestranol cause an increase of triglycerides, this is not to be expected when natural estrogens, such as estradiol, estradiol valerate, estrone and estriol, are administered in the usual substitutive doses [NIELSEN *et al.*, 1976]. Conjugated estrogens cause an increase in triglycerides only in doses above 1.25 mg [NOTELOVITZ and SOUTHWOOD,

1974; PYÖRÄLÄ, 1976]. Estriol and estradiol may in some cases even decrease elevated triglyceride levels [PROOST *et al.*, 1969; LEBECH and BOGGAARD, 1974; BOLTON, 1976; BORGLIN and STALAND, 1975; STROHFELDT, 1977; LARSSON-COHN, 1976]. When an increase in triglycerides is found, and when it is probably caused by the estrogens, the dose of this hormone should be reduced. A changing of the preparation from, for instance, ethinyl estradiol or conjugated estrogens to estriol, estriol succinate, dehydroepiandrosterone or estradiol, is sometimes effective. Additionally it is sometimes advisable to give a diet with a low triglyceride content or to prescribe medicaments which decrease the triglycerides. Furthermore, the parenteral administration of estradiol esters, oral or injectable estriol, or the local administration of estrogen ointments can be considered. Progestagens of the norethisterone type may additionally lower cholesterol and triglyceride values [NIELSEN *et al.*, 1976].

Cholestasis, Cholecystitis, Cholelithiasis

All steroid hormones, especially estrogens and above all the C-17-alkylated estrogens, may lead to a cholestasis in predisposed patients [MUELLER and KAPPAS, 1964]. When estrogens are to be administered in high doses and for a long time the patient should be asked if she has a history of liver or gallbladder diseases [Boston Collaborative Drug Surveillance Program, 1974]. The risk factors for gallbladder disease are given in table VI. To prevent cholestasis the dose of estrogens should be low. If the patient is taking also other medicaments which may cause cholestasis, these may have to to be withdrawn. When cholestasis occurs during treament, estrogens should be discontinued. If the therapy is resumed later, alkylated

Table VI. Risk factors for gallbladder disease

Positive family history
Earlier gallbladder complaints or disease
Overweight, high fat diet
Hyperlipidaemia, diabetes, hypertension
Multiparity
Hypothyroidism
Diseases of stomach and intestines
Intake of medicaments which may cause cholestasis (e. g., tranquillizers)

estrogens should be avoided and only low doses of natural estrogens or, if possible, local treatment with estrogens is recommended.

Pancreatitis

The risk factors for pancreatitis are shown in table VII. Estrogens aggravate a pancreatitis, and are contra-indicated in cases where a pancreatitis shows up in the anamnesis [Boston Collaborative Drug Surveillance Program, 1974]. When a pancreatitis occurs during therapy, estrogens and other steroids, as well as alcohol, should be withdrawn.

Table VII. Risk factors for pancreatitis

Alcohol abuse
Diabetes
Infections
Hyperlipidaemia
Hyperparathyroidism
Auto-immune diseases
Glucocorticosteroid intake

Table VIII. Risk factors for the occurrence of thrombo-embolism

Positive family history
Vessel diseases, anomalies and operations
Heavy smoking
Multiparity, age > 40 years
Severe varicosis, ulcus cruris
Overweight
Diabetes
Hyperlipidaemia
Hypertension
Cardiac or circulatory insufficiency
Anaemia, sickle cell disease
Blood group A, B, or AB, as opposed to 0
Blood and fluid loss
Tissue trauma
Immobilization
Infections
Intake of medicaments such as corticosteroids

Table IX. Risk factors for the occurrence of coronary thrombosis, myocardial infarction and cerebro-vascular accidents

Positive family history
Hypertension
Obesity
Diabetes
Hyperlipidaemia
Angina pectoris
Heavy smoking
Early menopause without estrogen substitution
Vessel anomalies
Severe stress

Table X. Risk factors for endometrial cancer

Positive family history of cancer
Individual cancer history
Other tumours (genital, mammary, intestinal)
Genetic and ontogenetic anomalies (Turner, Stein-Leventhal syndromes)
Long-lasting amenorrhoea, sterility, anovulation, nulli-, oligo-, late parity
Late menopause
Frequent curettages, endometrial hyperplasia, X-ray castration
Obesity (especially in tall women)
Diabetes
Hirsutism

Thrombosis and Thrombo-Embolism

Some estrogens enhance blood coagulation reactions, especially when given in higher doses and when the estrogen preparations are highly active and long-active. Oral estrogens (by direct action on the liver following gastrointestinal resorption) apparently influence the blood coagulation more than estrogens given parenterally. In women taking the contraceptive pill, a significant increase of deep venous thrombosis and embolies has been demonstrated. This might be related to an increase in factors VII, VIII, IX and X, of fibrinogen level, thrombocyte aggregation, a decrease of antithrombin III in interplay with plasminogen and the vessel wall fibrinolysis [Davies *et al.*, 1975; Poller, 1976; Bonnar, 1976]. No such increase has yet been observed when treating peri- or post-menopausal women with natural estrogens or with lower doses of the same estrogens. The statistics of the studies which have been done are open to criticism and certainly the number of patients was not as high as in hormonal contraception studies. There are few reports, such as the Boston Collaborative Drug Surveillance Program [1974], where no thrombosis and embolies have been seen in groups of post-menopausal patients treated for 5 years with natural estrogens.

Before estrogen treatment is prescribed, the patient should be asked if she or any member of her family has had a thrombosis, a thrombo-embolism, a coronary or a cerebral thrombosis. Risk factors for thrombo-embolism and for coronary thrombosis and myocardial infarction, are given in tables VIII and IX. Where there is an increased risk of deep venous thrombosis and thrombo-embolism and, nevertheless, steroid hormones are required, low doses and, especially, preparations which do not influence the blood coagulation should be given. Micronized estradiol in doses of 2–4 mg, 2 mg of estradiol valerate and estriol up to 6 mg/day do not influence the most important parameters of blood coagulation, such as prothrombin time, antithrombin III, factors II, VII, VIII, IX, and X, thrombocyte aggregation and vessel wall fibrinolysis [AMBRUS, 1976]. Medication on alternate days apparently diminishes the effect on coagulation parameters. Also, intramuscular injection of depot estrogens apparently exerts only insignificant effects on blood coagulation. The local application of estrogens should certainly be considered. The medicamental prophylaxis of thrombosis is very important. In cases of increased risk of thrombo-embolic complications estrogens should be withdrawn when an operation is planned or when long immobilisation is envisaged. When this is not done a peri-operative heparin therapy with low doses ($2 \times 5{,}000$ U/day) is recommended. This will lessen the risk of thrombo-embolism. In cases of increased risk the additional administration of salicylate and dipyridamole, together with the estrogen-progestagen medication, should decrease the thrombo-embolic risk. The prophylactic effects of physiotherapy and of special elastic stockings should also be pointed out. The danger of heavy smoking and the increased risk of belonging to certain blood groups (A, B, AB, factor S) is to be stressed [Editorial BMJ., 1970]. It has been claimed recently that appropriate screening especially by determining antithrombin III before treatment may result in the detection of those women who are particularly at risk of developing thrombo-embolic complications [AMBRUS, 1976].

Endometrial and Mammary Cancer

Many papers, both experimental and clinical, exist on the possible co-carcinogenic effects of estrogens. The fact that now, after nearly 50 years experience with estrogen therapy, we are still not sure as to whether or not estrogens play a role in carcinogenesis seems to me to indicate that no simple cause:effect relationship exists. When all the available evidence is considered, the opinion of BUTENANDT [1950] that estrogens are not directly carcinogenic still seems valid. It is possible, however, that under certain conditions and preliminaries they may exert an organotrophic 'facultative co-carcinogenic action'. This facultative co-carcinogenic action can, however, only come to realization, under the following conditions: (1) the estrogens have to be given in higher than normal doses; (2) they must be administered for a very long time; (3) the medication must be continuous, i. e., without interruption for a phase of proliferative rest, and (4) without the phasic addition of the natural estrogen antagonist progesterone. It seems questionable, however, that estrogens can lead to the manifestation of a cancer without there already being predisposing or real carcinogenic factors. These are shown in the figure 1.

The role of hormones seems to be a permissive one or to have a terrain effect, in that growth hormone, prolactin, androgens and estrogens are, dose dependent, either proliferatively (low or medium doses) or antiproliferatively effective (in high doses). Thus they may increase or decrease the likelihood of cancer development and may enhance or inhibit the tumour growth. The problem of carcinogenesis is very complex and is certainly not adequately considered, if it is judged only from the standpoint of estrogens or even only from that of hormones. The precarcinogenic and carcinogenic phases last for a very long time and over several escalating steps.

On the other hand, all experimental and clinical experience so far is in favour of the expectation that through a meticulous observation of the most important rules of hormone therapy the facultative co-carcinogenic effect of estrogens and other proliferative hormones (e. g. androgens) can be prevented, inhibited or even reversed. The evidence of many case/control and other retrospective studies points to the fact that a low dose and cyclical administration reduces the co-carcinogenic risk, and that the addition of a sufficient dose of progestagens may even reduce the risk [GREENBLATT, 1976; GAMBRELL, 1976]. The special risk factors for endometrial cancer are listed in table X. If several of these factors are present,

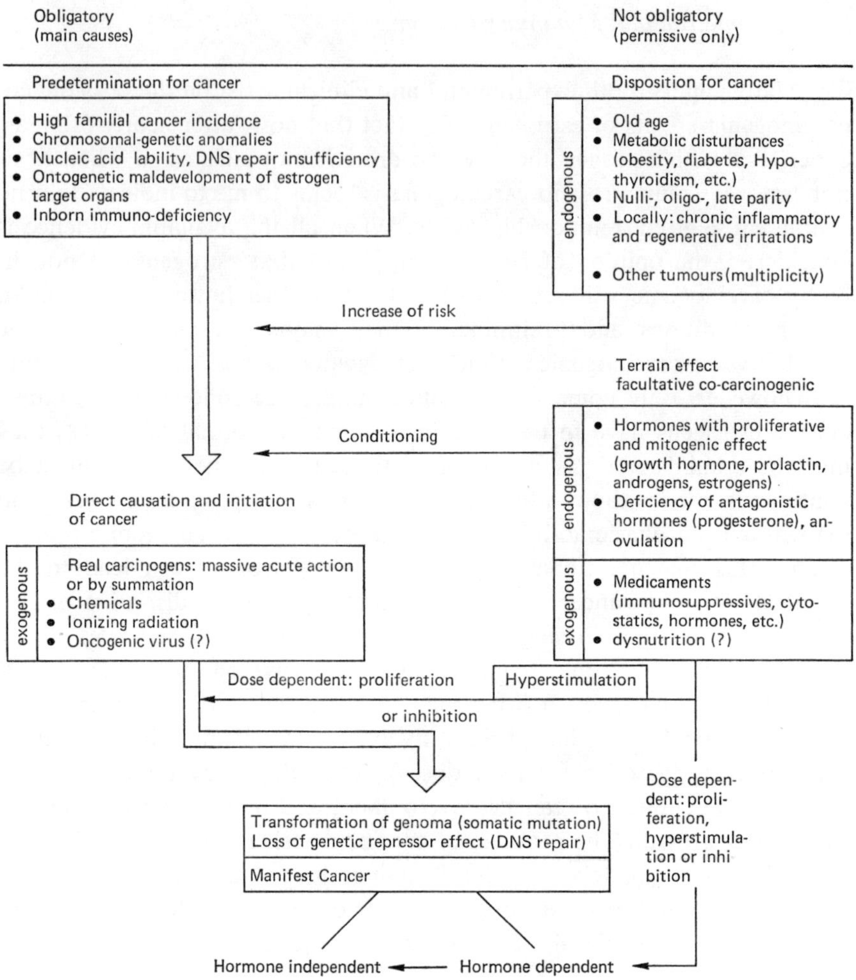

Fig. 1. Carcinogenic factors in endometrial and mammary cancer.

the risk of a cancerous disease by fate is high. I treat such patients with progestagens only, sometimes with additions of small cyclical doses of estriol. If the risk seems less, I give a cyclical treatment with low doses of natural estrogen and progestagen as a combined or sequential therapy (table X, XI). The patients are followed very closely and a gynaecological check-up is made after 6 months of therapy. A jet wash is done, and a suction curettage is considered before treatment and plans are made for repetition according to the requirements of the individual case.

Table XI. Risk factors for mammary cancer

Positive family history of benign or malign breast tumours
Cycle anomalies, anovulation
Sterility, nulli-, oligo-parity, late first pregnancy
Mastopathia
Early menarche, late menopause
Blood antigen S

The risk factors for mammary cancer are given in table XI. Again in cases of high risk, a medication of progestagen only or of small doses of androgens is recommended, especially if a mastopathy is present. As estriol is not co-carcinogenic in animal experiments, and can even reduce the development of metastases in small rodents treated with carcinogens, and causes less proliferation in the human female, its prescription is advocated for patients with a possible predisposition for breast changes [DRILL, 1960, 1965; BRECHER and WOTIZ, 1967; LEMON, 1970]. Estriol can be combined, if necessary, with dehydroepiandrosterone, testosterone, or a progestagen of the nortestosterone type or with anti-estrogenic properties.

In cases of a treated endometrial and mammary cancer with an apparently good prognosis, a progestagen can be given in the first 2 years for the treatment of severe climacteric complaints. The subjective relief is rather good with 5–10 mg/day of norethisterone acetate or lynestrenol. Alternatively, androgens (testosterone, dehydroepiandrosterone), sedatives, tranquillizers, vegetatives or clonidine (for hot flushes and sweating only) can be prescribed.

It is certainly not my intention to play down the cancer problem. The discussion concerning possible connections between estrogens and carcinogenesis should stimulate us to revise the indications, contra-indications, and risk factors, and to follow more closely the well-established rules for estrogen therapy. The most important rules to be observed are given in table XII. It should be mentioned perhaps that one of the advantages of long-term treatment and of a prophylactic therapy is that the patient is followed very closely by her physician. Thus, when unscheduled bleedings occur, a fractionated curettage can be done and, if it exists, a carcinoma of the endometrium, the cervix, or the mammary gland will be detected very early. This is a definite advantage, which seems to be suggested by the statistics of BURCH *et al.* [1975], which showed a reduction in the mortality of all cancers in their patients. Apparently it is a clinical fact

Table XII. Rules to be observed when prescribing estrogen therapy

The patient should be fully informed and give her consent to treatment
Never persuade a patient to accept a treatment she does not want
Consider indication
Pay regard to contra-indications and risk factors
Treat individually, not schematically
Use the lowest effective estrogen dose
Always treat cyclically (1 week's pause every 4th week), except perhaps when prescribing estriol
Add a progestagen 10 days per month, if feasible
Reduce the dose when side effects like leg cramps, oedema, or rapid weight gain occur, or when breast discomfort and cervical hypersecretion signalize overdosage or high sensitivity to estrogens
If indicated, consider treatment with estriol alone or the local application of estrogen
Make routine gynaecological and breast examinations and take a Papanicolaou smear every 6 months
In cases of atypical uterine bleedings perform a fractionated curettage
Consider hysterectomy (sooner than in non-treated women)

that in patients who receive estrogens the danger sign of abnormal bleeding occurs earlier than in non-treated cases, and that more often pre-stages are detected before real carcinomata occur. Moreover, when a carcinoma is detected in the hormone-treated patients, it is usually of the histological ripe form, which, having not penetrated into the muscular wall of the uterus, has a good prognosis with a cure rate of more than 90% [LAURITZEN, 1976].

My general rule with regard to estrogen treatment is never to prescribe it if the patient does not want it. When the therapy is to be given it is important that the patient be fully aware of the possible risks as well as of the advantages, and be willing to co-operate closely with her doctor. When this is done, and when the therapy is given according to the conditions listed above, it is my conviction that the advantages of the therapy outweigh any possible risks.

References

AMBRUS, J. L.: Estrogens and clotting factors; in VAN KEEP, GREENBLATT and ALBEAUX-FERNET Consensus on menopause research, p. 55 (MTP Press, Lancaster 1976).

BOLT, H. M. and BOLT, H. W.: Pharmacokinetics of mestranol in man in relation to its oestrogenic activity. Eur. J. clin. Pharmacol. *7:* 295 (1974).

BOLTON, C. H.: The effect of ethinylestradiol and conjugated equine estrogens on plasma lipids in oophorectomized women; in CAMPBELL The management of the menopause and post-menopausal years (MTP Press, Lancaster 1976).

BONNAR, J.: Effects of synthetic and natural estrogens on the coagulation system in post-menopausal women; in CAMPBELL The management of the menopause and post-menopausal years (MTP Press, Lancaster 1976).

BORGLIN, N. E. and STALAND, B.: Oral treatment of menopausal symptoms with natural oestrogens. Acta obstet. gynec. scand., suppl. 43, p. 1 (1975).

Boston Collaborative Surveillance Program: Surgically confirmed gallbladder disease, venous thrombo-embolism and breast tumors in relation to post-menopausal estrogen therapy. New Engl. J. Med. *15:* 290 (1974).

BOYD, G. S.: Oestrogens, cholesterol metabolism and atherosclerosis; in VAN KEEP and LAURITZEN Ageing and estrogens. Front. Hormone Res., vol. 2, pp. 74–97 (Karger, Basel 1973).

BRECHER, P. J. and WOTIZ, H. H.: Competition between estradiol and estriol for end organ receptor proteins. Steroids *9:* 431 (1967).

BURCH, J. C.; BYRD, B. F., and VAUGHN, W. K.: The effects of long-term estrogen administration to women following hysterectomy; in VAN KEEP and LAURITZEN Estrogens in the post-menopause. Front. Hormone Res., vol. 3, pp. 208–214 (Karger, Basel 1975).

BUTENANDT, A.: Östrogene und Krebs. Dt. med. Wschr. *5:* 75 (1950).

CRANE, M. G.; HARRIS, J. J., and WINSOR W.: Hypertension, oral contraceptive agents and conjugated estrogens. Ann. intern. med. *13:* 74 (1971).

DAVIES, T.; MCNICOL, G. P.; FIELDHOUSE, G.; GALLAGHER, J. C., and NORDIN, B. E. C.: A comparison of the effects of oestriol succinate and ethinylestradiol on blood coagulation, platelet function and fibrinolysis in post-menopausal women; in VAN KEEP and LAURITZEN Estrogens in the post-menopause. Front. Hormone Res., vol. 3, pp. 185–198 (Karger, Basel 1975).

DAXELMUELLER, L.; LAURITZEN, C., und HEILBRUNN, A.: Therapie des klimakterischen Syndroms mit Östriol-Succinat. Z. Therapie *12:* 14 (1974).

DRILL, V. A.: Physiology and pharmacology of steroids affecting tumor growth; in PINCUS and VOLLMER Biological activities of steroids in relation to cancer, p. 25 (Academic Press, New York 1960).

DRILL, V. A.: Relationship of administered estrogen, progesterone and oral contraceptives to cancer of breast and genital organs of women. Pacif. Med. Surg. *73:* 395 (1965).

Editorial: Smoking and oral contraceptives. Br. med. J. *iii:* 759 (1970).

EIFF, A. W. VON: Blood pressure and estrogens; in VAN KEEP and LAURITZEN Estrogens in the post-menopause. Front. Hormone Res., vol. 3, pp. 177–184 (Karger, Basel 1975).

ENGLUND, D. and JOHANSSON, E. D. B.: Pharmacokinetic and pharmacodynamic studies on oestradiol valerianate administered orally to post-menopausal women. Acta obstet. gynec. scand., suppl. 65, p. 27 (1977).

FOTHERBY, K.: Pharmacology of natural and synthetic estrogens; in CAMPBELL The management of the menopause and post-menopausal years, p. 87 (MTP Press, Lancaster 1976).

FURMAN, R. H.: Endocrine factors in atherosclerosis; in SCHETTLER and BOYD Atherosclerosis (Elsevier, Amsterdam 1969).

GAMBRELL, R. D., jr.: Estrogens, progestagens and endometrial cancer; in VAN KEEP, GREENBLATT and ALBEAUX-FERNET Consensus on menopause research (MTP Press, Lancaster 1976).

GREENBLATT, R. B.: Estrogens and endometrial cancer; in VAN KEEP, GREENBLATT and ALBEAUX-FERNET Consensus on menopause research (MTP Press, Lancaster 1976).

HEUSER, H. P. and STAEMMLER, H. J.: Histological investigation into the effect of oestriol succinate on the corpus uteri in post-menopausal women. Arzneimittel-Forsch. *23:* 558 (1973).

INMAN, W. H. W.; VESSEY, M. P., and WESTERHOLM, B.: Thrombo-embolic disease and the steroidal content of oral contraceptives. Br. med. J. *ii:* 203 (1970).

KALKOFF, R. K.: Effects of oral contraceptive agents and sex steroids on carbohydrate metabolism. A. Rev. Med. *23:* 429 (1972).

KOLB, K. H.: The metabolism of estradiol valerate. Med. Mitt. *28:* 16 (1967).

LARSSON-COHN, U.: Lipids and estrogens, in VAN KEEP, GREENBLATT and ALBEAUX-FERNET Consensus on menopause research, p. 51 (MTP Press, Lancaster 1976).

LAURITZEN, C.: Biologische Wirkungen einiger Östrogenmetabolite. 8. Symp. Dt. Ges. Endokrinologie, p. 233 (Springer, Berlin 1962).

LAURITZEN, C.: Oestrogens and endometrial cancer; in Clinics in obstetrics and gynaecology, vol. 4, No. 1, The menopause (Saunders, London 1976).

LEBECH, P. E. and BOGGAARD, B.: Serum lipids and antithrombin changes affected by synthetic and natural estrogen therapy; in GREENBLATT, MAHESH and DONOUGH The menopausal syndrome (Medcom, 1974).

LEMON, H. M.: Inhibition of rat mammary carcinogenesis by an impeded estrogen, estriol. Proc. Am. Ass. Cancer Res. *11:* 48 (1970).

LONGCOPE, C. and WILLIAMS, K. J. H.: The metabolism of estrogens in normal women after pulse injection of ^{3}H-estradiol and ^{3}H-estrone. J. clin. Endocr. Metab. *38:* 602 (1974).

MUELLER, M. N. and KAPPAS, A.: Estrogen pharmacology. The influence of estradiol on hepatic disposal of BSP in man. J. clin. Invest. *43:* 1905 (1964).

NEWTON, M. A.; SEALY, J. E.; LEDINGHAM, J. G., and LARAGH, J. H.: High blood pressure and oral contraceptives; changes in plasma renin, renin substrate and in aldosterone excretion. Am. J. Obstet. Gynec. *101:* 1037 (1968).

NIELSEN, F. H.; HONORE, E.; KRISTOFFERSEN, K.; SECHER, N. J., and THOMSEN-PEDERSEN, G.: Changes in the serum lipids during treatment with norgestrel, oestradiolvalerate or Cyclo-Progynova, p. 37. XIXth Nordic Congr. of Obstetrics and Gynecology, 1976.

NOTELOVITZ, M. and SOUTHWORD, B.: Metabolic effect of conjugated oestrogens (USP) on lipids and lipoproteins. S. Afr. med. J. *48:* 2552 (1974).

NOTELOVITZ, M.: The effect of natural estrogens on blood pressure and weight in post-menopausal women. S. Afr. med. J. *49:* 2251 (1975).

POLLER, L.: A double blind cross-over study with conjugated equine estrogens on blood clotting; in CAMPBELL The management of the menopause and post-menopausal years (MTP Press, Lancaster 1976).

PROOST, J.; DEMOL, R., and BARBIER, F.: Oestriolbehandeling van de menopauze en de post-menopauze. Bull. Soc. belge Gynéc. Obstét. *39:* 65 (1969).

PYÖRÄLÄ, T.: The effect of synthetic and natural estrogens on glucose tolerance, plasma insulin and lipid metabolism in post-menopausal women; in CAMPBELL The management of the menopause and post-menopausal years (MTP Press, Lancaster 1976).

RAURAMO, L.: Discussion in VAN KEEP and LAURITZEN. Estrogens in the post-menopause. Front. Hormone Res., vol. 3, p. 182 (Karger, Basel 1975).

Royal College of General Practitioners: Oral contraceptives and health (Pitman Medical, London 1974).

SPELLACY, W. N. and BIRK, S. A.: The effect of intrauterine devices, oral contraceptives, estrogens, and progestagens on blood pressure. Am. J. Obstet. Gynec. *112:* 912 (1972).

STROHFELDT, P.: Klinische Untersuchungen über Stoffwechselwirkungen von Östriolsuccinat. Arzneimittel-Forsch. *27:* 2006 (1977).

SYVALAHTI, E.; ERKOLA, R.; PUNNONEN, R., and RAURAMO, L.: Serum growth hormone, serum immunoreactive insulin and blood glucose. Response to intravenous tolbutamide in post-menopausal women with and without natural estrogen therapy. Acta pharmac. tox. *38:* 1 (1976).

UTIAN, W. H.: Definitive symptoms of post-menopause – Incorporating use of vaginal parabasal cell index; in VAN KEEP and LAURITZEN Estrogens in the post-menopause. Front. Hormone Res., vol. 3, pp. 74–93 (Karger, Basel 1975).

VELARDO, J. TH.: The anatomy and endocrine physiology of the female reproductive system; in VELARDO The endocrinology of reproduction (Oxford University Press, New York 1958).

Prof. C. LAURITZEN, Department of Gynaecology and Obstetrics, University of Ulm, Prittwitzstrasse 43, *D-7900 Ulm/Donau* (FRG)

Discussion

RAURAMO: I welcome this balanced and wise clinical approach. It must be accepted that estrogens, because they are such potent substances, can indeed have side effects. However, as they are produced naturally and are necessary for the normal functioning of the female body, they cannot in themselves be regarded as harmful. Even if, in very high dosages, they might be regarded as dangerous, we should not ignore their very real usefulness when properly administered in physiological quantities.

I think everyone here would agree that estrogen therapy properly used can be extremely beneficial around the time of the menopause and in the post-menopause, though it is clear that, as individuals, we differ considerably in our enthusiasm. I think too that we should stimulate further research into the effects and risks of the various estrogens, and that for the time being we should prescribe only the preparations, dosages and regimens which we know to be effective and safe.

Front. Hormone Res., vol. 5, pp. 248–260 (Karger, Basel 1978)

General Discussion

(Chairman: Prof. B. E. C. NORDIN)

NORDIN: Two days of papers and discussions have given us enough material to come now to a concluding session in which the benefits of estrogen therapy can be balanced against the risks. The first question I want to put forward is this: What are the reasons which prompt climacteric women to come and ask for medical help?

LEMON: In the United States women ask specifically for estrogen therapy because they want to stay young for ever.

RAKOFF: That *was* my impression too; in the sixties many women came because of that. At that time I often had difficulties in dissuading those for whom I regarded such therapy as unnecessary or even contraindicated. In the last 2 years – since 1975 – the picture has changed because many women have been scared by the publicity around potential risks. Now, more often than not, it is the hot flushes and the perspiration that induce them to come.

UTIAN: I have worked on the climacteric problems of women in several different countries, and have found there to be a great variety in the pictures that women present. There are, of course, always the typical complaints – the hot flushes, perspiration, atrophic vaginitis – but the rest of the picture is largely determined by social and cultural factors. The expectations of the woman play an important role, and in this respect she is often influenced by what she reads in the mass media, in magazines and so forth. It is the medical profession's responsibility to explain the true facts, the benefits *and* the risks, and we should not hesitate to use the mass media ourselves in order to get this message across. It is often a combination of fear, anxiety and ignorance which determines whether or not a woman asks for help.

VAN KEEP: Our figures refer only to European countries and, dating from 1969, are already somewhat old. We found then that around 40% of the 45- to 55-year-old women we interviewed had seen a doctor in connection with the menopause. About one third of the others said that they had not experienced any difficulties, but the rest, some 40%, said that they had not sought medical help even though they had experienced menopausal problems. Various reasons were given for this: some feared the gynaecological examination and others were afraid that the doctor would find something seriously wrong, but it became clear that a major reason was ignorance of the fact that something could be done to help them. It was interesting to see that the country in which the highest percentage of women had sought medical advice, West Germany, was also the one in which the most women knew that a special therapy existed. The study was done just after the explosion of publicity there on the 'stay young for ever' theme.

LAURITZEN: That particular wave of publicity really made women, and indeed some doctors, aware of the existence of treatment of climacteric symptoms and of estrogen therapy, but the quest for eternal youth has never involved many women in Germany. Most women come to ask for help for their typical climacteric complaints, such as hot flushes and sweating. A considerable number come because of nervous symptoms, i. e., depression, or a decrease in vitality: they no longer feel like themselves, they can no longer 'cope with the world'. These women need more help than mere pharmacotherapy. They need understanding, reassurance, and advice, but in these cases estrogen therapy will certainly make adaptation easier. Sexual problems appear to be quite common, but they are rarely voiced spontaneously. One has to probe in order to learn about them.

KELLER: In a rather conservative society such as Switzerland's, women do not often or easily ask for help in connection with sexual problems. Neither do many Swiss women see a connection between psychic or osteoporotic problems and the end of the fertile period. If they do ask for help, it is often not the gynaecologist who is approached first.

MCKAY HART: Patients coming to our clinic in Glasgow do so at their own request. They fall into three categories. Firstly, there are the women with flushes and sweats. These are usually working women or those who in other ways are often in situations where they are embarrassed by the thought that their flushes may be noticed by others. Other women report that the sweats come at night and are disturbing both for themselves and for their husbands. The second group comprises women with psychic problems, and, thirdly, there are women whose relatives have had fractures of the hip and who think that they themselves may be subject to the same sort of problem. I have no doubt that it is women in the first and third groups that we can help most.

VAN DER DOES: More and more women over 35 are now coming for regular medical check-ups, many without spontaneously voiced complaints. It is often when taking the medical history of these women, especially those who are peri-menopausal, that one uncovers a whole variety of problems which may justify therapy.

VAN KEEP: There is quite a large group of women who come with the whole gamut of complaints. When one asks about these, one finds that many of them have already existed for years, but now that she also has hot flushes, the whole lot have been labelled 'climacteric', both by the patient herself and by her general practitioner. This sort of women has completely unrealistic expectations of estrogen therapy. When such patients are carefully screened, not only by the gynaecologist but also by the internist, pathology, sometimes long standing, often is found.

STUDD: Women often accept the hot flush as an inevitable part of ageing, but what really brings them to us is that they feel 'awful'. They cannot concentrate, cannot do their work properly, their family life is a shambles and so on. When they finally come to the menopause clinic they have already been treated by their general practitioners with all imaginable tranquillizers and sedatives, therapies which I think have no place whatsoever in the treatment of the climacteric. If a patient who has been well and been in command of her work and family life, suddenly falls in the leap through no apparent reason, she is most probably going to benefit from estrogen therapy. Some 25 % of all women between 45 and 55 are on tranquillizers and only 3 % are on estrogens [VAN KEEP, P. A.: Soziologische Aspekte des Klimakteriums. Arch. Gynäk. *224:* 370–377 (1977)]. I myself see about 5 patients a week who are on psychotrophic drugs, sometimes two or three types at the same time. Provided they fall in the category I have described, they are completely off these drugs after just 3 months of hormone therapy.

UTIAN: The mental tonic effect of estrogens has now been confirmed in several studies. It is an aspect of therapy that we should not underestimate.

KOPERA: When I reviewed the literature on the effects of estrogens on psychic functions for our 1972 workshop, I was struck by the fact that so little seemed to be known [in Ageing and Estrogens. Front. Hormone Res., vol. 2, pp. 118–133 (Karger, Basel 1973)]. I am aware that several publications have appeared since, but it still seems to me that strangely little work is being done in this field. I would have thought that the psychological aspects of the climacteric are of such great practical importance that many more researchers would have been actively engaged in studying these complex phenomena.

RAURAMO: The psychological effects of estrogens are very important if we want to give women an easy change from the fertile period to the post-menopause. I cannot help feeling that the reason for so few women being found in leading positions is to be at least partly explained by the mental unbalance in these years around the time of the menopause. It is around the age of 50 that men take the final step to the top, a step that women with equal intellectual capacities rarely take. I know that many aspects are involved, but the climacteric may well be an important one.

BAULIEU: In this connection I think that it is important that female physicians be involved in menopause research. I have the feeling that they are underrepresented. They certainly are at this meeting. I feel that their presence is desirable especially when we are discussing the psychological aspects of the climacteric and what women expect from therapy.

RAKOFF: Women have their expectations, but so do I as their physician. I have to think also in terms of preventive medecine. I must be aware which processes can be prevented and which cannot. Women also need information about this.

NEVES E CASTRO: If we come to he conclusion that the value of estrogen therapy goes beyond its value in treating climacteric complaints – and I think we must accept that it has value as a preventive medicine – we should give up our passive attitude. We should not just wait for patients to come to us, but should educate them and encourage them to come for therapy. Furthermore, if we think there are women who will in later life suffer from the effects of estrogen deficiency, we should try to develop means of identifying these women so that preventive therapy may be given.

MACK: We are perhaps a little pompous in discussing the reasons why women come to the physician. Medicine is, in a way, a service industry, and informed women come to the doctor because they wish to obtain a service, provision of reliable advice. They may come for other less appropriate services, in which case we have to ask for help from the behavioural scientists. They may ask the wrong questions because they are poorly informed, and in this case, we have an educational job to do. But in actual fact we ourselves are by no means certain that we have a therapeutic service to offer either for short-term relief of symptoms or for long-term prevention of complications. As far as the short-term is concerned, about which we are fairly knowledgeable, we must inform women clearly what the facts are and what we consider the pros and cons of the treament to be. Long-term therapy is more problematical: few women, I think, come to us at present with this in mind. If we truly believe it to be of value then, I agree, we should go out and bring the patients in. But how sure are we of the long-term benefits? Do we know, for instance, to what degree we are able to prevent life-threatening fractures in old age by giving estrogens? It seems to me that we are really very ignorant about the effects of estrogen therapy on morbidity and mortality.

NORDIN: Let me now sum up this part of the discussion. Hot flushes and sweats are the most common reasons for women coming to ask for medical help. Next comes the loss of confidence, which can be worded in many different ways: problems in the woman's social life, in her family life, at work, and so on. Third on the list of spontaneously mentioned problems come the sexual problems: loss of libido, loss of interest, dryness of the vagina.

When one looks at the elicited responses, the order is more likely to be: flushes first, sexual problems second, and the 'loss of confidence' problems third.

We all agree that there is a recognizable and definable climacteric entity which is worth treating and which reacts readily to estrogens. We do not know – at least, not yet – any tool to enable us to predict which woman is likely to respond to therapy and which is not. Clinical experience may, however, be of help; if the psychological problems and the sexual problems have been present for years and have only recently been complicated by hot flushes, the result of estrogen therapy is likely to be poor. If, however, the onset of these problems has more or less coincided with the first occurrence of hot flushes the chances that the therapy will succeed are very good.

As far as prevention is concerned the situation is a little different. Patients who come to the menopause clinic usually have nothing wrong with their bones, but it is now possible to identify the fast bone losers: they are the women with a high fasting urine calcium, a high hydroxyproline excretion, a low vaginal (or urinary) smear maturation value, low plasma estrone and androstenedione levels, and a thin skin. This does not mean, however, that we are now in a position to introduce large scale screening programmes and give estrogens to all those found likely to develop osteoporosis later in life. The reason why this cannot be done is that doubts have been voiced about the safety of long-term estrogen therapy, and whilst it is acceptable to give a risk carrying therapy to a patient who has a specific treatable syndrome, it is quite another thing to encourage people to use the same therapy merely because their urinary calcium is high and the use of the therapy may prevent, say, a wrist fracture in some 10 years' time. It is clear that before the widespread use of estrogen therapy can be advocated as a preventive medicine the question of the risks must be sorted out.

Now that we have defined the climacteric syndrome which responds well to estrogen therapy, let us move on to the treatment of this syndrome. I do not think there is much value in our all describing our preferred treatment schedule, but it would be interesting to hear your thoughts on one point, that of how long treatment should be continued.

HASPELS: I continue treatment, as long as the patient has complaints. I then explain the benefits of continuing beyond that stage, and find, in fact, that most of my patients choose to do so.

LAURITZEN: In my experience the matter tends to resolve itself. When the symptoms no longer occur most women either ask to be taken off the therapy or simply do not return to the clinic. This may be after a few months or after some years. A small number, usually the more intelligent, well-informed, health-conscious women ask for the therapy to be continued; these are very often women who have been using oral contraceptives for a long time and who are used to having regular medical checkups. In these cases I continue the therapy for as long as it makes sense to do so. With advancing age, reasons to discontinue gradually come to the fore. Long-term treatment is certainly necessary in early castrated women, in females with long-standing climacteric complaints, in patients with symptomatic atrophic changes of the genitalia, and in women with osteoporosis.

VAN DER DOES: I treat as long as symptoms are present, but I try to decrease the dose gradually. If I want to see if it is possible to stop completely I usually do this just before the patient goes on holiday. The change in life's rhythm helps them to do without estrogens.

RAKOFF: If you administer cyclically you can see the moment approach when you can discontinue. If no complaints occur in the tablet-free intervals the patient may be able to do without tablets completely.

NORDIN: So it seems that we feel that, as far as climacteric complaints are concerned, there is a natural point for discontinuing therapy. This natural point is often determined by the patient herself.

My next question is: Has the time come when the administration of a progestagen in addition to estrogen therapy for the treatment of the climacteric syndrome has become mandatory?

UTIAN: As far as I am concerned, no. It may be beneficial for some patients, but I think that more work is needed before we can be as categorical as to say that it should be given to all patients.

HASPELS: The advantages of adding a progestagen are not proven. I have found that many patients who get this additional progestagen feel less well – they feel depressed, bloated, and report a loss of libido. I feel that estrogen therapy, under good medical supervision, is sufficient.

LAURITZEN: It is strange, though, that people report this depressive mood on combined therapy because if one gives progestagens alone, even in quite elevated dosages, such symptoms are not reported.

RAKOFF: A disadvantage of the combined therapy is that it causes 'menstruations' to restart, and many women are not willing to accept this.

LAURITZEN: It depends on how the patient is told about it. If it is explained to her that such bleedings are an extra safeguard against hyperproliferation of the endometrium and that this form of therapy may prevent undesirable changes in the breasts it is easier to accept. I only combine estrogen therapy with a progestagen, however, if there is a reason to do so; if, for instance, the patient is a high risk patient.

VAN KEEP: It depends on the dose and on which estrogen one uses. If there is no risk of proliferation there is also less reason to give a progestagen.

VOOIJS: I would like to mention one point, and that is that whilst progestagens have an effect predominantly in the superficial layers of the endometrium, estrogens may have a marked one in the basal layers as well. If one adds a progestagen for a while and then withdraws both hormones, one sometimes sees that an existing hyperplasia remains, and sometimes even increases, whilst the upper layers appear quite normal.

STUDD: I am undecided about the addition of a progestagen. Half of our patients on oral therapy are now receiving courses of 5–7 days of norethisterone, but like Professor HASPELS, we have found that many of these women are complaining of pre-menstrual tension, depression, and breast discomfort. Of course, all the patients who have an intact uterus and who are on estrogen implants receive 5–7 days of norethisterone. I would like to hear from Dr. MACK whether or not he has found that cyclical therapy and the addition of a progestagen has any influence on the relative risk.

MACK: The relative risk was indeed lower with cyclical therapy, but the proportion of women on estrogens *and* progestagens was too small to permit conclusions to be drawn. A study has recently been done with Provera; it will undoubtedly attract a lot of publicity when it is published, but unfortunately the numbers are too small and it cannot be interpreted.

LAURITZEN: One more point about the induced menstrual bleedings. We have found that on prolonged combined therapy the withdrawal bleeding becomes less and less abundant over the years and sometimes disappear completely. It is as if the target organ decreases in responsiveness with ageing.

NORDIN: We have found that the incidence of bleeding on the 3 weeks on, 1 week off regimen is a question of the dose of the estrogen and the age of the woman. The larger the dose and the younger the woman the more likely she is to bleed. We too have noticed that patients on a sequential type regimen feel less well than on pure estrogen. I have the impression that we are not sure how much progestagen is needed. Some people use 1 mg of norethisterone, others use 5 mg.

LAURITZEN: The progestagen dose should be high enough to ensure transformation of a proliferation or even hyper-proliferated endometrium, i. e., at least 10 mg norethisterone acetate, 3 mg norgestrel, or 50 mg 17α-hydroxyprogesterone derivatives. The duration of medication should be at least 8–10 days.

NEVES E CASTRO: A practical approach seems to me to administer just progestagens – and no estrogens – to women in the pre-menopause who have an irregular cycle or an amenorrhoea. I prescribe a 7-day course of a progestagen from the 18th to 26th day of each cycle. As long as they bleed, I consider that they are producing sufficient estrogens. When no bleeding occurs their menopause has come. At that time I discuss with the patients whether or not a pure estrogen therapy is indicated. I think that in this way I prevent irregular bleedings and endometrial hyperplasia. To a large extent I also prevent psychic problems and, perhaps, at least to some extent, vasomotor symptoms.

NORDIN: I want to conclude this part of the discussion with a few remarks. There seems to be no relation between hot flushes and the estrogen deficiency syndrome which leads to, amongst other things, osteoporosis. In my opinion the two, the occurrence of the hot flushes and the estrogen deficiency syndrome, tend to be mutually exclusive: the osteoporotic patient, generally speaking, has never had hot flushes. There is no relation between hormonal levels and symptomatology. I should perhaps make an exception for Dr. STUDD's point that in pre-menopausal women one may use the FSH levels to distinguish between a genuine climacteric syndrome and psycho-neurosis, but after the menopause all women have raised gonadotrophins, there is a wide range of estrone and androstenedione levels and a wide range of symptoms.

It may be that the climacteric syndrome is an episode in the woman's life when she goes through a particular band of hormonal levels, and probably fluctuates within that band. The faster she gets through that band, the more quickly she gets into the estrogen-deficient state, and the more quickly she is through the climacteric syndrome. This may take anything from a few weeks to several years. Thus it may be that long continuing hot flushes are a good sign as far as a woman's prognosis in terms of estrogen deficiency is concerned. A further piece of evidence for this is that if women with the typical signs of estrogen deficiency osteoporosis, high hydroxyproline, low estrogen levels, a low maturation value – but with no hot flushes are treated with estrogens, hot flushes sometimes suddenly occur in the beginning of therapy. It really seems that we are dealing with two different problems: the climacteric syndrome and the estrogen deficiency syndrome.

We now have to come to the risk aspects of estrogen therapy. In this meeting we have been primarily informed about relative risks and not so much about absolute risks. It seems to me that the absolute risk of having an endometrial carcinoma is very small for the woman who is not taking estrogens and that estrogen therapy, although it may increase this risk, does not really involve a substantial *absolute* risk to the woman.

MACK: It is the relative risk, though, which is pertinent when a clinician is deciding whether or not to prescribe estrogens. It is not the absolute risk.

NORDIN: But the absolute risk matters to the woman.

MACK: What matters to the woman is the absolute excess burden she is taking on. If a woman of, let us say, 60 years of age has an absolute risk of 50/100,000 and the relative risk of estrogen use would be 8, the figure goes up to 400/100,000. The attributable risk is then 400 minus 50: that is 350/100,000 woman-years. It is this increase in risk that she takes on and that has to be balanced against the benefits measured in similar terms per 100,000 woman-years. In breast cancer, where the base line absolute risk is closer to 200/100,000 woman-years, a relative risk of 2 adds 200; that is a lot, though the relative risk is low.

NORDIN: As a clinician I am more interested in the absolute risk without and with estrogen therapy than in the relative and attributable risks, because these latter concepts are difficult to appreciate. The mortality of appendectomy is said to be three times higher in non-teaching hospitals than in teaching hospitals, but people's behaviour is not influenced by this fact (if it is a fact) because the absolute danger is still so very low.

MACK: It might also be because they are poorly informed. If I had to choose I would prefer the one third risk. I think that it is your responsibility to inform the patient based on what you believe to be the quantitative phenomena, of the risk and of the benefit.

UTIAN: If the risk for a woman at the age of 50 to have endometrial carcinoma is 1/1,000 per annum and the risk does not change with age – which I presume it does – then the absolute risk for her will be about 25/1,000 in the 25 years she still has to live. If the relative risk of estrogen taking would be 8, can I say that her chances will be 8 times 25, that is 200?

MACK: Yes, that is right.

NORDIN: I have the impression that clinicians are emotionally biased against such figures because they do not want to believe that a therapy they find effective and useful can be dangerous. Conversely, I think that epidemiologists must admit that it is marvellous to be able to show data that tell a positive story, and perhaps they are biased the other way.

MACK: I really object to that. I do not accept your statement that the epidemiologist is biased *per se*. In the past I have with equal satisfaction disagreed with previous epidemiological studies.

UTIAN: I too think that there is too much intellectual investment by workers – clinicians and epidemiologists – to accept, a priori, a major degree of bias. We are attempting to be honest. I think, therefore, that we should use the information that we have. Dr. MACK and Dr. RAURAMO have both shown us their data and there is a discrepancy between the two reports: there is an enermous increase in relative risk by estrogen use in the community that Dr. MACK studied and there is no increase in the relative risk in Finland. Can we explain this discrepancy? Dr. MACK has shown us, and I am sure that his data are correct, the risk at its worst. The explanation might be found in the different therapeutic approaches. In the first case there were uncontrolled high dosages, continuous therapy with potent estrogens, in the second case there were low dosages, discontinuously, of weaker estrogens administered under close supervision of the doctor.

THIJSSEN: I think, however, that Dr. RAURAMO's interpretation of his data is still open to criticism. He did not give relative incidence figures for endometrial cancer in Finland, but gave absolute figures. His control group seems biased too, because as many as 19% of all women in this group were on estrogens.

NORDIN: It is also said that 45% of all post-menopausal American women are on estrogens, yet in SMITH's control group only 15% were on estrogens against 45% of his endometrial cancer patients. One has to accept such case control studies as they stand.

RAURAMO: Indeed, in the population, from which both the cases and the controls were taken about 15–20% uses estrogens. I gave absolute figures for endometrial cancer to show that, although the population as a whole is on the decrease, there is no decrease in the frequency of this cancer. I must say too that the final data of our study still have to be analysed, and we shall in the final report give the relative age adjusted incidence of endometrial cancer during the last years. These figures are not available at the present time.

NORDIN: One point that strikes me is that the clinicians here today, who must be able to amass a considerable number of woman-years on estrogen therapy between them, have not found evidence of an increase in endometrial cancer. Is this because of the way in which we manage our patients? Is it, perhaps, because we have all been using cyclical treatment?

MACK: First of all, we found that cyclical treatment with low doses gave a relatively lower estimated risk of 2½. I cannot comment on your statement, though. I should have to look at all those cases in great detail. I remember when GRAY in Kentucky, a great believer in estrogen therapy, was convinced that with his therapy he was preventing breast cancer; he challenged MACMAHON to study his material. When HOOVER, MACMAHON's associate, analyzed the data he not only found that estrogens were not preventing breast cancer but that there was actually (and frighteningly) a suggestion of a positive association. Now I do not even want to suggest that something like that might happen if the patient material of all of you were screened by an epidemiologist, but I must say that there is insufficient ground to accept your complacency on this point.

UTIAN: I myself would hesitate to draw any statistical conclusion on the basis of my own clinical observations. I have seen recently a case of endometrial cancer and remember a few cases of breast cancer.

LEMON: I have a question in this respect. If there were a minor increased risk of breast cancer associated with estrogen therapy this would have enormous consequences because the base line risk is already so high. Prof. LAURITZEN did not mention in his second paper the need for a mammography prior to estrogen therapy. I think this, in addition to a breast examination, would be most desirable. If the patient later developes breast cancer you might blame the estrogen therapy for accelerating the growth of an undetected breast cancer. This may even have legal consequences if one has not taken due precautions to rule out such individuals from treatment. I have testified in this type of case in which a settlement was paid by the defending gynaecologist.

LAURITZEN: It is not usual in Germany to do a mammography prior to the institution of estrogen therapy, not even in risk patients. Perhaps it should be done. It would be, of course, most interesting, from the scientific point of view to have mammographies – and endometrial biopsies too – taken of all patients before therapy. However, mammographies are more often made of patients on estrogens than of comparable groups of patients without estrogen substitution. This may also be true of other diagnostic measures such as curettages, vaginal smears, rectal examinations, and others.

NORDIN: Are there case control studies implicating estrogens in breast cancer? Is there a known association between endometrial cancer and breast cancer?

LEMON: As far as your second question is concerned, yes, I think so.

DE WAARD: There are several papers stating this. There is one from MACMAHON and AUSTIN [Cancer *23:* 275–280 (1969)] and another from BAILAR [Cancer *16:* 842–852 (1963)].

LAURITZEN: We must not forget that there is a certain multiplicity of cancer in general. Patients with a carcinoma of the uterus have a greater than expected incidence of carcinoma of breast, colon, rectum, oral cavity, and skin [MOERTEL, C. G.: Multiple primary malignant neoplasms (Springer, Berlin 1966)].

NORDIN: There is also a relation between breast cancer and a late menopause; the later the menopause the greater the risk [STOLL, B. A.: in Endocrine therapy in malignant disease, p. 119 (Saunders, London 1972)].

LEMON: In the study of BURCH and BYRD [Estrogens in the postmenopause. Front. Hormone Res., vol. 3, pp. 208–214 (Karger, Basel 1975)] the incidences of breast cancer in their patients slightly exceeded that of the expected ratio. At least half of their patients were castrated, so one would have expected a reduced risk.

LAURITZEN: Oophorectomy will reduce mammary cancer risk only in patients less than 40 years old. The slight increase in breast cancer incidence in the data of BURCH and BYRD may also result from cancer multiplicity as stated above.

MACK: As to the influence of the age at menopause, the problem in studying estrogens and breast cancer is that it is difficult to eliminate this confounding factor: the later the menopause, the fewer years at risk of using estrogens. Various investigators have tried, but no one has really yet succeeded in a definitive study.

NORDIN: I have another provocative question. Of the eight studies on the relation between estrogen use and the risk of endometrial cancer, about five have come out with negative findings and three with positive ones. Have the negative ones all been poorly designed?

MACK: Even if this statistic were correct (and I do not accept that it is) it would be ridiculous for anyone to base his conclusions on such a comparison. He should evaluate each study and see whether or not it meets his criteria for a well-designed study. I refer to my paper here, and must state that the studies with a negative outcome were less well designed than the others.

NORDIN: I have to sum up here. The epidemiologists at this meeting have made a very strong case for there being some connection between estrogen therapy and an increased risk of endometrial cancer. As far as breast cancer is concerned the question is still open. Many clinicians present find it difficult to reconcile these data of the epidemiologists with their own clinical experience.

We concluded that careful patient supervision is required, together with proper action if danger signals appear. However, estrogen therapy is extremely valuable in the climacteric and may also be useful in the postmenopausal period for both the treatment and prevention of estrogen deficiency – of which osteoporosis is the most striking manifestation at present – and that we are not prepared to abandon it unless the evidence against it becomes more convincing than it is at the moment.